Advances in Applied General Equilibrium Modeling

Series editors

This series has a companion series in *SpringerBriefs in Applied General Equilibrium Modeling*.

The series publishes advances in the theory, application, parameterisation and computation of applied general equilibrium (AGE) models. AGE analysis is now an essential input in many countries to the discussion of a wide range of economic topics relevant to public policy. This reflects the capacity of AGE models to carry extensive economic detail, their flexibility in accommodating new policy-relevant theory and data, and their capacity to project economic outcomes for a large number of macroeconomic and microeconomic variables.

Topics in AGE modeling addressed by the series include: macroeconomic forecasting and adjustment; public finance; economic growth; monetary policy and financial markets; environmental policy; energy policy; income distribution and inequality; global modeling; country-specific modeling; regional modeling; economic effects of natural disasters and other catastrophic events; productivity; demography; foreign direct investment; economic development; model solution algorithms and software; and topics in estimation, calibration and validation.

AGE applications are increasingly multi-disciplinary, spanning inputs from such diverse fields as engineering, behavioral psychology, energy modeling, land use modeling, demography, and climate modeling. The series allows for the comprehensive documentation and careful exposition of not only the AGE models themselves, but also the inter-disciplinary inputs to the modeling, and the interactions between each.

For AGE modelers, the series provides a format supporting: clear exposition of data work, attention to the theoretical modeling of relevant policy detail, and thorough discussion of simulation results. This aids both academic and policy readerships.

Academic readers will appreciate: the capacity to see details of the full complexity of relevant components of model equation systems; comprehensive documentation of data manipulation algorithms; supporting analysis and discussion of model input and closure assumptions; and careful discussion of results grounded in AGE theory, data and closure assumptions.

Policy readers will appreciate: a format that supports the reporting of the comprehensive set of model outputs of interest to policy makers; discussion of elements of the theory and data that exert a heavy influence on research findings; and nuanced and qualified discussion of the policy implications of AGE research.

More information about this series at http://www.springer.com/series/13860

Glyn Wittwer
Editor

Economy-Wide Modeling of Water at Regional and Global Scales

Springer

Editor
Glyn Wittwer
Centre of Policy Studies
Victoria University
Melbourne, VIC, Australia

ISSN 2520-8268 ISSN 2520-8276 (electronic)
Advances in Applied General Equilibrium Modeling
ISBN 978-981-13-6100-5 ISBN 978-981-13-6101-2 (eBook)
https://doi.org/10.1007/978-981-13-6101-2

Library of Congress Control Number: 2018967425

This Springer imprint is published by the registered company Springer Nature Singapore Pte Ltd.
The registered company address is: 152 Beach Road, #21-01/04 Gateway East, Singapore 189721, Singapore

Preface

An edited volume on water modeling was the suggestion of one of the series editors of Advances in Applied General Equilibrium Modeling, James Giesecke. This arose during the 19th Annual Conference on Global Economic Analysis held in Washington DC in June 2016.

The objective of this volume is to bring material together from contributors whose perspectives differ widely. This is because the characteristics of water resources within the nations analyzed within the volume vary widely. Moreover, different stages of economic development present different problems concerning water resource management. More generally, various models that embed water into an applied general equilibrium (AGE) framework are examples of the versatility and policy relevance of AGE models.

It is appropriate that the founder of the Global Trade Analysis Project (GTAP), Tom Hertel, and his colleagues at Purdue University are among the contributors. Their perspectives using variants of the GTAP model are global, national, and sub-national. It is also appropriate that Mark Horridge is co-author of one chapter. He is the creator of The Enormous Regional Model (TERM) used in analysis in two chapters of the volume covering three nations at a sub-national level. Various models outlined in this volume cover specific nations from five continents.

The volume may provide a useful reference for other modelers who are grappling with national accounts data and recognizing the need for auxiliary water accounts to proceed with model development. From a policy perspective, climate change is a pervasive issue. Structural changes in economies are altering water needs. We may anticipate a growing economic literature on AGE water analysis as the relevance of the framework becomes more widely recognized.

Melbourne, Australia
November 2018

Glyn Wittwer

Contents

Editor and Contributors

About the Editor

Prof. Glyn Wittwer is a regional dynamic CGE modeling expert. He has played a major role with Mark Horridge in developing databases for TERM versions in several countries. He edited the Springer volumes *Economic Modeling of Water* (2012) and *Multi-regional Dynamic General Equilibrium Modeling of the U.S. Economy* (2017) and contributed the majority of chapters in both volumes. Glyn is a GTAP Research Fellow from 2017 to 2020. He has extensive consulting experience. His list of projects includes dynamic, multi-regional CGE modeling in Australia, China, and the USA. These include modeling of the impacts of major dam and transport projects, drought and water trading, flood, hypothetical plant disease scenarios, productivity scenarios, wine tax scenarios, major mine construction projects, industry closures, and the aftermath of civil disruption.

Contributors

Philip D. Adams Centre of Policy Studies, Victoria University, Melbourne, Australia

Roy Brouwer Department of Environmental Economics, Institute for Environmental Studies, Vrije Universiteit Amsterdam, Amsterdam, The Netherlands Department of Economics and The Water Institute, University of Waterloo, Waterloo, Canada

Santiago Vianna Cuadra Empresa Brasileira de Pesquisa Agropecuária (Brazilian Agricultural Research Corporation), Campinas, SP, Brazil

Angel dos Santos Fachinelli Ferrarini Universidade Estadual de Maringa (State University of Maringa), Maringá, PR, Brazil

Joaquim Bento de Souza Ferreira Filho Department of Economics, Business and Sociology, School of Agriculture "Luiz de Queiroz", University of Sao Paulo, Piracicaba, SP, Brazil

Tom Hertel Department of Agricultural Economics, Purdue University, West Lafayette, IN, USA

Mark Horridge Centre of Policy Studies, Victoria University, Melbourne, VIC, Australia

Joachim C. Hunink Deltares, Delft, The Netherlands

N. Kilimani Department of Economics, Makerere University, Kampala, Uganda

Jason F. L. Koopman Wageningen Economic Research, The Hague, The Netherlands

Onno Kuik Department of Environmental Economics, Institute for Environmental Studies, Vrije Universiteit Amsterdam, Amsterdam, The Netherlands

Jing Liu Department of Agricultural Economics, Purdue University, West Lafayette, IN, USA

Ruslana Rachel Palatnik Department of Economics and Management, and Sustainable Environmental and Economic Development Research Center (SEED), Yezreel Valley College, Afula, Israel Natural Resources and Environmental Research Center (NRERC), University of Haifa, Haifa, Israel

Elizabeth L. Roos Centre of Policy Studies, Victoria University, Melbourne, Australia

Roberto Roson Ca' Foscari University, Venice, Italy
IEFE Bocconi University, Milan, Italy
Loyola Andalusia University, Seville, Spain

Farzad Taheripour Department of Agricultural Economics, Purdue University, West Lafayette, IN, USA

Richard S. J. Tol Department of Environmental Economics, Institute for Environmental Studies, Vrije Universiteit Amsterdam, Amsterdam, The Netherlands Department of Economics, University of Sussex, Falmer, UK
Department of Spatial Economics, Vrije Universiteit Amsterdam, Amsterdam, The Netherlands

Marnix P. van der Vat Deltares, Delft, The Netherlands

Jan H. van Heerden Department of Economics, University of Pretoria, Pretoria, South Africa

Daniel de Castro Victoria Empresa Brasileira de Pesquisa Agropecuária (Brazilian Agricultural Research Corporation), Campinas, SP, Brazil

Glyn Wittwer Center of Policy Studies, Victoria University, Melbourne, Australia

List of Figures

List of Tables

Chapter 1
The Contexts for Water Modeling

Glyn Wittwer

Abstract Rapid economic growth in previously less developed nations and prolonged droughts in many nations are stressing existing water resources. Diminishing water resources are being diverted increasingly to meet burgeoning urban demands. Economics has a role to play in providing guidelines to enhance the sustainability of water resources. Water pricing is important in improving resource allocation. Economic models can help place policies in context. Economic adjustment to water stress may alter patterns of production and global trade.

Keywords Drought impacts · Water reforms · Farm factor mobility · Water trading

1.1 Background

Irrigated agriculture has been the lifeline of some civilizations for millennia. In northwestern China and other parts of central Asia, the *karez* system of wells and tunnels has given farmers and communities access to gravity-fed water for over two thousand years. In Oman, the *aflaj* system of irrigation may have been distributing water since around 2500 BC.[1] The *aflaj* system entails flexibility, in terms of separation of water and land ownership, and the possibility of water trading alongside of which some present-day institutional arrangements in Western nations appear primitive. The civilizations of Egypt managed a more obvious source, the Nile, even earlier.[2] In Sichuan province, China, the Dujiangyan irrigation system, created around 250 BC, diverts water for irrigation, drains sediment, controls floods and regulates flows without the use of dams. Modifications and enlargements during later dynasties

[1] See http://whc.unesco.org/en/list/1207/, accessed 10 September 2018.

[2] http://www.waterhistory.org/histories/nile/t1.html#photo1, accessed 5 September 2018.

G. Wittwer (✉)
Centre of Policy Studies, Victoria University, Melbourne 8001, Australia
e-mail: glyn.wittwer@vu.edu.au

G. Wittwer (ed.), *Economy-Wide Modeling of Water at Regional and Global Scales*, Advances in Applied General Equilibrium Modeling,
https://doi.org/10.1007/978-981-13-6101-2_1

have resulted in an irrigation area that now covers almost 700,000 ha.[3] UNESCO has accorded World Heritage status to the best sites in the *aflaj* system and the Dujiangyan system.

It is when we move from the ancient to the modern that the danger signals appear in irrigation water management. A couple of decades of rapid mining expansion in China's Xinjiang province have lowered the water table, threatening the ancient *karez* system. In Oman, farmers who have lost patience with the old ways of managing water have started sinking wells to draw water with a similar consequence. These are just two examples of modern practices that are disrupting age-old irrigation systems (Amri et al. 2014).

Many modern irrigation schemes have failed or are failing. Short-term economic goals are often to blame. One villain is food self-sufficiency: such a drive in Saudi Arabia resulted in a peak in Saudi Arabian wheat production of over 4 million tonnes in the early 1990s. At its peak, arid Saudi Arabia became a major wheat exporter. There is no more stark an example of production and trade that is contrary to comparative advantage. But such a departure from economic trading fundamentals could not last: the depletion of aquifers, which consisted of fossil water with no possibility of recharge, resulted in an almost complete cessation of wheat production by 2015.

Saudi Arabia's example of groundwater depletion is extreme, but the threat of depletion also faces farmland in northern India, Pakistan, north-eastern China, South Africa, Iran, countries of northern Africa and large parts of the USA, with the most rapid depletion in California. Wada et al. (2010) analyzed global groundwater stress in 2000. The stress has worsened substantially since then. Contributing to this has been rapid economic growth in India and China, the drought in California from 2012 to 2015, rapid depletion of the Ogallala aquifer in central U.S. states and the millennium drought in Australia. A recurring theme among contributors in this volume is responding to prolonged drought.

Irrigation schemes used to facilitate economic development can have dire consequences. The most extreme case is that of the Soviet era diversions from the Amu Darya and Syr Darya rivers to irrigate cotton fields in the desert in Uzbekistan. The volume of water in the Aral Sea is now a few percent of what is was in 1960, prior to irrigation development.[4] The development has adversely affected lifestyles around the sea, destroying the sea's fishing industry. Boats are stranded dozens, if not hundreds of kilometers from the nearest shoreline. Without the moderating influence of the sea, summers are hotter and winters are colder than they were prior to the development of the irrigation scheme. Toxic dust storms exacerbate the difficulties faced by remaining inhabitants.

Fortunately for inhabitants on the Kazakh side of the border, a World Bank project to build a 12 km dyke, plus other works to improve flows in the Syr Darya river, have raised water levels in the North Aral Sea since 2005. This has revived the fishing

[3]http://whc.unesco.org/en/list/1001, accessed 5 September 2018.

[4]See http://earthobservatory.nasa.gov/IOTD/view.php?id=84437, accessed 5 September 2018.

industry in some areas (Chen 2018). It serves as a reminder that interventions to improve environmental and economic outcomes are possible.

1.2 The Role of Economics

The track record of many modern irrigation developments is discouraging. In some countries, there have been attempts to combine engineering solutions with other approaches, including legal and economic reforms. In Australia, for example, reforms in the 1990s separated water titles from land titles. This led to the establishment of water exchanges: water trading has evolved into a routine matter. Such trading played a significant role in helping farmers in the Murray-Darling Basin respond to the millennium drought. But economically reasonable policies, such as water buybacks by the Australian government to deal with unsustainable allocations in the Murray-Darling Basin under the 2007 Water Act, have met with political resistance. Far less cost-effective irrigation infrastructure upgrades have been more appealing to many politicians, representing the dominance of engineering at any cost, and without credible quantification of the benefits, over economics.

The role of economics is not as an all-encompassing cure. Rather, given the challenges that governments, industries and communities around the world face, economic tools have a role in the response to rising water scarcity. Economics is concerned with optimality. Full environmental restoration of a river basin, for example, is likely to be sub-optimal, given that it may come at the cost of foregone output and jobs. Moreover, resources dedicated towards restoration of one resource may divert resources from environmental restoration of other natural resources. Yet partial restoration such as the above example concerning the North Aral Sea, may yield substantial benefits for relatively modest costs.

Metering of water usage and pricing of water may improve environmental outcomes. In particular, water pricing may slow groundwater depletion. Price mechanisms are important in dealing with scarcity. As water scarcity worsens, an underuse of such mechanisms will lead to a greater misallocation of water. To enable legal and economic mechanisms to take effect, engineering works must be in place. For example, water trading between regions cannot take place unless their water supplies are connected.

Politically acceptable water management decisions often take precedence over economically justifiable decisions. For example, the mining town of Broken Hill in outback New South Wales, Australia, suffered water shortages around 2010 and beyond, as the water supply from the Medindee Lakes failed due to ongoing dry conditions. Ever anxious for a final solution, the NSW government approved construction of a $500 million pipeline to connect the town to the more reliable water supply of the southern Murray-Darling Basin.[5] This is equivalent to a cost of $25,000

[5]See https://www.industry.nsw.gov.au/water/plans-programs/infrastructure-programs/broken-hill-pipeline, accessed 10 September 2018.

per person in Broken Hill. Prior to the 2018 completion of the pipeline, the town turned to bore water when the supplies failed. The pipeline is being built despite a rigorous scientific study concluding that using managed aquifer recharge, presumably at a fraction of the cost of the pipeline, could enhance the security of the town's water supply (Lawrie et al. 2012).

Over the past 40 years, rapid population growth in China's major cities, combined with rising incomes have escalated urban water demands. At the same time, climate change has made water supplies less reliable. Jobs, land and water are moving from agricultural to urban activities. China's South-to-North-Water-Diversion (SNWD) Project is an extreme version of a grand engineering scheme to redress water shortages in China's north. The north is relatively water scarce. The eastern component of SNWD connects Jiangsu, Anhui, Shandong, Hebei and Tianjin to the Yangtze River, using in part the ancient Grand Canal. The central route of SNWD, which become operational in 2014, links the Yangtze with northern cities, now supplying one third of Beijing's water needs (Barnett et al. 2015). The cost of the project exceeds US$80 billion. 340,000 people in total from the cities of Shiyan and Nayang in Hebei were relocated as part of the central route. Barnett et al. (2015), who are geographers not economists, characterize the Chinese approach as serving the interests of those who supply water, namely the state-owned hydro-engineering firms. Decentralized approaches that could be adopted in the cities of China's north include augmenting existing water supplies with rainwater catchment on buildings, increased uptake of water recycling, separation of water according to quality and more stringent pollution controls.

Far from assuring the water supply of China's north, the SNWD Project has in some instances along the Grand Canal (eastern route) connected formerly unpolluted segments of water with polluted segments. Barnett et al. (2015) note that the maintenance costs of SNWD may be much higher than original estimates. This implies that transferred water may be prohibitively expensive for farmers.

In the case of farmers, Barnett et al. (2015) attribute water shortages to inefficient water use practices in farming and a lack of water storage. The authors conclude that improved water management could do much to extend water availability in northern China. The authors speculate quite reasonably that "the SNWD project might well mark the nadir of big-engineering solutions to China's water problems" (p. 27).

This may not yet to be the case. A detailed discussion of hydropower is beyond the scope of this volume. But as noted by Barnett et al. (2015), hydropower authorities are among the apparent winners from the construction of massive engineering schemes within China to supply water. Controversial hydropower constructions include the Zangmu Dam in Tibet. It became operational in October 2015, disrupting rivers that supply Bangladesh and India. Wong (2016) notes that as some dam projects in Tibet have proceeded, others have halted. The future of dam construction on Nu River, which passes from Tibet through Myanmar before emptying into the Andaman Sea, is uncertain, given growing fears of disruption to tourism, fisheries and local communities.

1.3 The Role of Economic Models and Outline of This Volume

There are many models of water allocation. Hydrologists, engineers or economists may have been constructed these. They may concentrate on very short-term delivery requirements or optimal economic allocation over a longer period. The latter models may include environmental accounts. Research might examine, for example, the marginal impacts on farm output of realizing a particular environmental objective. Griffith (2012) discusses some of the many models of water allocation and outlines the various objectives of different types of water models.

What are the advantages of a CGE approach? An obvious feature of CGE models is that they include price responses. The shadow price of water (i.e., the value of marginal product of water) may vary widely between seasons as water availability changes. The average product of water differs markedly between different farm activities and fluctuates with output prices. Already, this hints at the idea that relatively rigid forms of production and water allocations may not be optimal, as water at a point in time may be far more valuable in one activity than another. Constraints on allocating water are not only physical. If land and water rights are bundled together, this limits the opportunities to trade water in order to improve water allocations.

Hertel and Liu in Chap. 2 provide a detailed review of the literature on water, water scarcity, sectoral activity and economic growth. They identify the opportunities and challenges faced by practitioners who wish to include water use in a CGE framework. The chapter addresses agricultural water consumption, water use in energy production, and demands for water by households, industry and services. Finally, it discusses water supply and allocation.

The chapter concludes that there are relatively few instances in which water scarcity has significantly slowed the long-term rate of national economic growth. Many opportunities for water efficiency improvements abound in electricity generation, production of industrial steam, residential consumption, and irrigated agriculture.

Hertel and Liu note that although water is essential in most economic activities, its substitutability with other inputs varies widely between economic activities. The aim of an economic model with water embedded in it is to consider reallocation of water using market mechanisms. For example, there are interactions between market distortions in agricultural and misallocation of water. The authors grapple with how modelers should treat water in different economic uses. Agriculture may require special treatment, and may include land-water bundles and farm factor mobility. Hertel and Liu note the possibility of building water into a household production function. Modelers could treat consumer durables as a form of capital stock. An implication is that investments in household appliances within a model with this feature would impact on water usage.

The problem usually associated with constructing CGE models is that they rely on national accounts data. They may lack spatial or sectoral resolution. Typically, published input-output tables that provide the foundation for CGE databases include few farm sectors. But a growing number of CGE models are appearing which deal with spatial and sectoral resolution. In the spatial dimension, the TERM (The Enormous Regional Model) methodology devised by Mark Horridge of the Centre of the Policy Studies (see Horridge et al. (2005) and Horridge (2012)) has revolutionized CGE modeling by enabling small region representation with full CGE detail. In the sectoral dimension, farm census data enable the CGE practitioner to disaggregate farm outputs across small regions.

Since farming is often a combination of dry-land and irrigated technologies, such disaggregation may not be sufficient for policy-relevant modeling. Further modifications are required to model irrigation water usage. For a start, there may be a need to split farm outputs into irrigated and dry-land technologies. Other modifications may include the addition of water accounts, a distinction between annual and perennial crops enabled by different degrees of factor mobility and water trading possibilities (see Chap. 5).

In Australia, the TERM-H2O model which concentrated on regions of the Murray-Darling Basin helped put into perspective the markedly different contributions of drought and water buybacks (introduced under the 2007 Act) to regional economic impacts. The TERM-H2O framework enables the practitioner to add a great deal of a detail to a multi-regional national model. Dynamic modeling may also estimate the marginal impacts of policy in the context of ongoing structural change. Table 5.1 show how much the observed price of water varies according to seasonal conditions. With water trading and mature markets, the trading price resembles the value of the marginal product of water in irrigation activities.

Chap. 5 also covers modeling of the prolonged California drought which started around 2012. The model represents the Central Valley regions at the county level. Despite some limited water trades with the modeling, there are marked differences in water price movements in each region, indicating that further water trading would have enhanced allocative efficiency in response to drought.

More challenging is modifying a global CGE model to deal with irrigation water. Since economists regard water trading as a mechanism for improving water allocations, allowing water trading possibilities in a model may enhance its analytical power. The difficulty in a global model concerns aggregation. Regions and sectors must be disaggregated sufficiently that water trading within the model remains geographically plausible. The GTAP model (Hertel 1997) provides a pragmatic platform for global representation with its disaggregation of agricultural sectors beyond what is found in most national input-output tables. Liu et al. (2016 and Chap. 3) have added sub-national region-specific farm industries (based on agro-ecological zones or AEZs) to national regions in a modified version of the global model called GTAP-BIO-W. This approach enables water trading between irrigators in a specific sub-national region, thereby confining such trades to realistic possibilities. In addition, Liu et al. include some of the theoretical elaborations of the national TERM-H2O model in the global framework. The advantage of a detailed CGE model is that it can depict varying degrees of factor and input mobility, including water, thereby enhancing realism. For example, in response to extreme water scarcity in one AEZ, farmers may be able to purchase feed inputs from other regions so as to maintain livestock herds.

Roson (Chap. 4) examines the impact that economic growth between now and 2050 will have on water demands. Given the unsustainable exploitation of water resources globally at present, we cannot expect that water demand will grow with an accommodating increase in supply. Rather, adjustment will occur in several ways. Technological change will reduce water requirements per unit of output. Minimum tillage practices have enhanced moisture retention and soil conservation. Biotechnology and other technologies are improving drought resistance and enhancing yields. Rising water scarcity will induce innovation that results in further water savings. The composition of output will change so as reflect changes in water scarcity. Desalination and wastewater treatment will play an increasing role. The composition of international trade will change, so that there is a change in the pattern of virtual water trading.

Country-by-country studies reveal that the hydrology, geography and economic development in each nation presents its own set of management issues. In turn, food security remains a policy concern in some nations either due to deficient infrastructure, regional politics and climate change. Uganda is water-abundant yet its citizens suffer from limited access to potable water. Brazil has stewardship over a significant share of the world's water supply, but is so vast that water scarcity varies markedly between regions and between cities. Israel is land scarce and water scarce. And even in the water-abundant Netherlands, climate changes looms as a major policy concern in future water and land management.

In Chap. 6, van Heerden et al. examine water policy issues in Uganda. The nation presents a case of ostensible water abundance. A different picture emerges once account is taken of Uganda's water sharing arrangements with other countries, the nation's relatively high population density and its limited sanitation infrastructure. The transport infrastructure in landlocked Uganda is also deficient. This implies that food security is a bigger concern for Uganda than for a nation with a greater capacity to trade. Liu et al. (2014) emphasize the role of trade to buffer water shortfalls but this possibility remains impractical in Uganda. Water efficiency gains in cropping therefore are a major concern as the nation's population continues to grow rapidly.

Brazil's water availability represents at least 12% of all the fresh water in the world. Ferrarini et al. (Chap. 7) examine regional impacts on the Brazilian economy of increasing irrigated agriculture. The chapter covers impacts on agricultural productivity and water demand, and the balance between water demand and water availability. The northeast region of Brazil is more prone to droughts than the rest of the country, and may be the most vulnerable to climate change. The multi-regional CGE model of Brazil used in this analysis distinguishes between irrigated and rain-fed agriculture. The basis of the simulations detailed in this chapter is the Brazilian National Water Resources Plan. Modeling draws implications for water sufficiency in each region from this. Climate change scenarios follow, in which Ferrarini et al. assume that northeast Brazil's water resources diminish.

Koopman et al. (Chap. 8) examine water issues in The Netherlands. Substantial investments in water infrastructure are used in this delta nation to prevent flooding, facilitate shipping transport, irrigate farmland and ensure the health of polder lands and nature. Although the Netherlands is a wet country, there are still times when irrigators cannot access water, to the detriment of agricultural production. The authors note that water is not privately owned, not always scarce, and not always allocated according to market principles. One prediction of the climate change scenarios is for the Netherlands to have hotter and drier summers and a substantial drop in summer river discharge. Modeling examines the impact of adaptation responses, including further investments in irrigation infrastructure and the re-allocation of water resources among competing uses.

Israel is in the unique position of being heavily reliant on desalination as a source of potable water despite being an oil-importing nation. Desalination accounts for a growing share of Israel's water needs. In the case of potable supply, the share is around one quarter. In Chap. 9, Palatnik uses the Israeli Computable General Equilibrium model (IGEM) to estimate the impacts of desalination on the Israeli economy. The model depicts five water types, and accounts for constraints associated with crop salinity-tolerance and food-safety regulations. Prolonged drought has contributed to desalinated water's marginal value exceeding its marginal supply cost.

1.4 Inferences from This Volume

Economic modelers face a spatial dilemma concerning water allocation issues. The relative scarcity of water at any time may vary widely from one watershed region to the next. This points to the importance of a local dimension in modeling. Yet the implications of worsening water scarcity are global. This means that rising water scarcity at the local level may imply different impacts from those at the global level.

Economists frequently counter doomsday predictions of resource depletion or environmental destruction with reference to induced innovation and productivity gains. Not all technology applied to water usage concerns direct improvements in efficiency in water use, be they in agriculture or elsewhere. Some improvements may arise from better use of water availability and meteorological data and medium-term meteorological forecasting. One example is that La Nina and El Nino meteorological events affect each side of the Pacific Ocean. These result in abnormally wet periods on one side of the Pacific and abnormally dry periods on the other at the same time. This natural cycle is one cause of fluctuations in water availability. Water users need to improve adaptations to and anticipation of changing water scarcity before plunging into a crisis. Worsening seasonal fluctuations due to climate change that seasonal fluctuations are hastening the need for adaptation strategies. In California, growers of perennials cope with declines in surface water availability by drawing more on groundwater. This makes sustainability of groundwater drawings a management issue. In Australia's Murray-Darling Basin, water trading enables perennial producers to purchase water from other irrigators when scarcity worsens.

The contributions of Liu et al. (Chap. 3) and Roson (Chap. 4) to modeling of water are in the global context. For a region or a nation, changing global scarcity may imply that reduced productivity in a region is offset partly by either a terms-of-trade swing or a change in the composition of farm outputs.

In the country-specific studies in this volume, the authors provide considerable background detail on the geographic and economic circumstances of each region. Such information will help the reader understand the issues that the modelers within each study explore. For example, without acknowledging underdeveloped transport infrastructure in Uganda, an economist may be tempted to conclude that policy concerns with food security within the national are misguided. In the case of Israel, analysts may conclude that desalination would be less necessary if the nation were less intent on maintaining the agricultural sector. Palatnik in Chap. 9 models the interaction between water use and economic activity with five different types of water. Fig. 9.1 indicates that little potable water is used in agriculture, dispelling the notion that agriculture is driving the demand for desalination.

This volume brings together different examples of global, national and small region models. Data are becoming more widely available. Modelers are improving the depiction of water in the economy in their models. The links between such modeling and policy implementation may not always be apparent. However, climate change, population growth and income growth are imposing increasing strain on water resources. This is likely to increase the demand for modeling to inform policy makers, and raises the potential benefits that may arise from modeling of water.

References

Amri S, Ghafri A, Rhaman N (2014) Water management of Falaj Al Khatmain in Sultanate of Oman. J Earth Sci Eng 4:127–133

Barnett J, Rogers S, Webber M, Finlayson B, Wang M (2015) Sustainability: transfer project cannot meet China's water needs. Nature 527:295–297

Chen D (2018) The country that brought a sea back to life. http://www.bbc.com/future/story/20180719-how-kazakhstan-brought-the-aral-sea-back-to-life. Accessed 5 Sept 2018

Griffith M (2012) Water resources modeling: a review. In: Wittwer G (ed) Economic modeling of water: the Australian CGE experience. Springer, Dordrecht, pp 59–77

Hertel T (ed.) (1997) Global trade analysis: modeling and applications. Cambridge University Press, New York

Horridge (2012) The TERM model and its database. In: Wittwer G (ed) Economic modeling of water: the Australian CGE experience. Springer, Dordrecht, pp 13–36

Horridge M, Madden J, Wittwer G (2005) Using a highly disaggregated multi-regional single-country model to analyse the impacts of the 2002–03 drought on Australia. J Policy Model 27:285–308

Lawrie KC, Brodie RS, Tan KP, Gibson D, Magee J, Clarke JDA, Halas L, Gow L, Somerville P, Apps HE, Christensen NB, Brodie RC, Abraham J, Smith M, Page D, Dillon P, Vanderzalm J, Miotlinski K, Hostetler S, Davis A, Ley-Cooper AY, Schoning G, Barry K, Levett, K (2012) Securing Broken Hill's water supply: assessment of conjunctive water supply options involving managed aquifer recharge and/or groundwater extraction at Menindee Lakes. Geoscience Australia. https://d28rz98at9flks.cloudfront.net/73819/Rec2012_011.pdf. Accessed 10 Sept 2018

Liu J, Hertel TW, Taheripour F, Zhu T, Ringler C (2014) International trade buffers the impact of future irrigation shortfalls. Glob Environ Change 29:22–31

Wada Y, van Beek L, van Kempen C, Reckman J, Vasak S, Bierkens M (2010) Global depletion of groundwater resources Geophysical Research Letters. https://doi.org/10.1029/2010GL044571
Wong E (2016) China's last wild river carries conflicting environmental hopes. New York Times, 18 June

Prof. Glyn Wittwer is a regional dynamic CGE modeling expert. He has played a major role with Mark Horridge in developing databases for TERM versions in several countries. He edited the Springer volumes *Economic Modeling of Water* (2012) and *Multi-regional Dynamic General Equilibrium Modeling of the U.S. Economy* (2017), and contributed the majority of chapters in both volumes. Glyn is a GTAP Research Fellow from 2017 to 2020. He has extensive consulting experience. His list of projects includes dynamic, multi-regional CGE modeling in Australia, China and the United States. These includes modeling of the impacts of major dam and transport projects, drought and water trading, flood, hypothetical plant disease scenarios, productivity scenarios, wine tax scenarios, major mine construction projects, industry closures and the aftermath of civil disruption.

Chapter 2
Implications of Water Scarcity for Economic Growth

Tom Hertel and Jing Liu

Abstract This chapter provides background information useful for a quantitative global assessment of the impact of water scarcity on growth using a Computable General Equilibrium (CGE) model. It provides a detailed review of the literature on water, water scarcity, sectoral activity and economic growth, and identifies the possibilities and bottlenecks in incorporating water use into a CGE framework. It covers water use in agriculture, energy production, households, industry and services. Finally, it discusses water supply and allocation. There are ample opportunities for conserving water across its various uses. Economic incentives would hasten water efficiency gains.

Keywords Water use · Water scarcity · Economic growth · CGE modeling

2.1 Economic Modelling of Water

The OECD Environmental Outlook Baseline projects future global water demand to increase by 55% between 2000 and 2050 and this is expected to make water one of the most fiercely contested resources on the face of the planet (OECD 2013) . The purpose of this chapter is to provide background information useful for quantitative assessment of the impact of water scarcity on the economy and its growth using a multi-region multi-sector Computable General Equilibrium (CGE) model. Before embarking on a detailed review of the literature on water, water scarcity, sectoral activity and economic growth, it is useful to revisit several strands of economic theory to put this issue in context. We explore two different aggregate economic models, each of which treats water in a different way. In so doing, we are able to

T. Hertel · J. Liu (✉)
Department of Agricultural Economics, Purdue University, 403 W State Street, West Lafayette, IN 47907-2056, USA
e-mail: liu207@purdue.edu

T. Hertel
e-mail: hertel@purdue.edu

G. Wittwer (ed.), *Economy-Wide Modeling of Water at Regional and Global Scales*, Advances in Applied General Equilibrium Modeling,
https://doi.org/10.1007/978-981-13-6101-2_2

shed light on several dimensions of water's impacts on economic growth which will need to be taken into account as we move forward with the incorporation of water into a CGE framework.

2.1.1 Aggregate Production Function Approach

Thinking about water as an input (W) into a national production function, we can write GDP (denoted as y) as a function of water and a composite input comprising both physical and human capital (K): $y = f(K,W)$. For simplicity let us assume that this production function is of the constant elasticity of substitution (CES) form with a CES parameter denoted by σ As society accumulates additional capital, we expect the capital/water ratio: K/W to rise, thereby inducing relative scarcity of water – assuming that the overall supply of water is limited by the hydrological cycle. In such an economy, the value of σ is critical to the potential for long run growth in this economy. If firms and households can take advantage of the increasingly abundant capital to invest in water-saving or reuse technologies, and if these technologies are sufficiently effective, then growth will proceed unabated. This is indicative of an economy with a large value of σ. Indeed, provided $\sigma > 1$, even in the absence of water-saving technological change, the share of water in GDP will diminish over time as the economy becomes more and more water-efficient through capital-water substitution. On the other hand, in an economy where output per gallon of water cannot be increased via capital investment ($\sigma = 0$), growth will be curtailed if water supplies cannot be increased through additional capital investment. Therefore, from an empirical point of view, it is important to estimate, at the sectoral level, the equilibrium substitutability between other inputs and water. This will include both the direct elasticity of substitution in the production function as well the response of sectoral output and input adjustments in the wake of increased water scarcity.

In practice the economy comprises many sectors, each with many different end uses for water. Water can also be processed and reused in many cases. Therefore, an economy-wide estimate of σ needs to reflect the possibility of such reuse. In addition, it must reflect not only the ability to become more efficient in specific end uses, but also the possibility of eliminating some end uses altogether. The economy-wide estimate of σ must also incorporate the potential to substitute away from products produced by particularly water-intensive technologies. These types of intra- and inter-sector substitution relationships are difficult to capture in a single aggregate economic model. Thus, the goal of multisector Computable General Equilibrium (CGE) models is to capture these effects. By including within the model the potential for technical substitution and innovations at a disaggregated level, as well as the potential to substitute away from water intensive intermediate and final goods, CGE models allow for an accurate assessment of the economy-wide potential for substitution of capital and other inputs for water. This is why CGE models will be a focal point of this survey of water scarcity and economic growth.

2.1.2 Water and Growth in a CGE Framework

A natural way to think about economic growth within a CGE framework is to track the per capita utility of a representative household in the economy. From a policy perspective, changes in utility are typically translated into monetary terms (e.g. $US) using the concept of equivalent variation (EV), or the change in real income associated with a reduction in water availability. EV is likely to be affected via a number of different channels. In this section we explore these channels in some detail, as they will determine the ultimate impact of water scarcity on economic growth and welfare.

$$EV_s = (\psi_s)\left\{\begin{array}{r}\sum_{i=1}^{E}(PE_{is}QE_{is}(\mathrm{d}QE_{is}/QE_{is}))\\ +\sum_{i=1}^{E}\sum_{j=1}^{J}(\tau_{Eijs}PFE_{ijs}\mathrm{d}QFE_{ijs})\\ +\sum_{i=1}^{N}(\tau_{Ois}PD_{is}\mathrm{d}QO_{is})\\ +\sum_{i=1}^{N}(\tau_{CDis}PD_{is}\mathrm{d}QD_{is})\\ +\sum_{i=1}^{N}(\tau_{CMis}PM_{is}\mathrm{d}QM_{is})\\ +\sum_{i=1}^{N}\sum_{r=1}^{R}(\tau_{Mirs}PCIF_{irs}\mathrm{d}QMS_{irs})\\ +\sum_{i=1}^{N}\sum_{r=1}^{R}(QMS_{irs}\mathrm{d}PFOB_{irs})\\ -\sum_{i=1}^{N}\sum_{r=1}^{R}(QMS_{irs}\mathrm{d}PCIF_{irs})\end{array}\right\} \tag{2.1}$$

Technical Preliminaries: Eq. (2.1) provides a stylized decomposition of regional economic welfare in the context of a global CGE model (Huff and Hertel 2001). In this equation EVs refers to the change in welfare in region *s,* and *r* denotes a trading partner region which is the source of imports into our focus region, *s*. ψ_s is a scaling factor which is normalized to one initially, but changes as a function of the marginal cost of utility in the presence of non-homothetic preferences (McDougall 2003). The subscripts *i* refer to produced commodities, of which there are *N* in total. As we will see below, in most CGE models, this is where municipal water shows up since the municipal water supply is provided by a public utility using scarce resources. Water can also appear as an endowment, alongside capital, labour and land, of which there are E endowments in total, each potentially employed in any of the J sectors.

The first way in which water scarcity affects welfare in this economy is the most obvious, direct channel—namely there is less water available for use. In the case of water endowments, this is captured by the first term in brackets ($PE_{is}QE_{is}(dQE_{is}/QE_{is})$ which reflects the current valuation of water's contribution to the economy. This is based on the 'shadow price' of water (i.e. *Endowment i*) in region s (PE_{is}), and its quantity (QE_{is}), thereupon multiplying this valuation of water by the proportional change in its availability, (dQE_{is}/QE_{is}).[1] So if the contribution of water to the regional economy is $1 billion, and there is a 10% reduction in its availability, then a first-order guess at the welfare cost would be 0.10*$1 bill. = $100 mill. This is just a first-order estimate, since any reduction in available water will affect the marginal value product of water. It will also interact with other features of the economy – a point to which we will come momentarily.

In a market-based model (we will discuss the problem presented by the absence of water markets shortly), the valuation of water is embedded in the market price, which is expected to rise with increasing scarcity. Whether or not the proportionate rise in price exceeds the proportionate decline in water availability can be related directly back to the economy-wide elasticity of substitution discussed above. If $\sigma > 1$, then price will rise less than quantity, and the value pre-multiplying the proportional change in water availability to the economy will fall. On the other hand, if water is essential to household and firm production processes ($\sigma > 1$), then the price will rise faster than the quantity reduction and the valuation of water in the economy will rise as water becomes more scarce. In this case, the penalty for the first 1% cut in water will be less than that which applies when cumulative reductions have reached 10%. The empirical literature on the price elasticity of demand for water discussed throughout this survey provides support for the hypothesis that, in many cases, we observe $\sigma > 1$—at least at the level of individual firms and households.

This direct impact of water scarcity on per capita regional welfare notwithstanding, there are quite a number of other ways in which water scarcity can affect welfare in region s indirectly. The first is via reallocation of water. As we have seen with the ongoing drought in California, water scarcity can result in significant shifts in different sectors' claims on the diminished water resources. When the marginal value product of water in different uses differs greatly within an economy, there is considerable potential for 'second best' efficiency effects from such reallocations. Consider the situation portrayed in Fig. 2.1 in which a pre-determined amount of water withdrawals, $\overline{W}$, is allocated between sectors A and B, as reflected by the initial allocations: W_A and W_B. This pattern of water allocation reflects the presence of an implicit subsidy on water use in sector A—or equivalently – a tax on Endowment Water use in sector B of region s: τ_{EWBs}—since the marginal value product of water in A (read off the A^*a segment) is lower than that in B (read off the B^*b segment) at

[1] This expression offers a local approximation to EV, for large changes, the price and quantity levels must also change. The ensuing numerical integration is what allows us to compute the sources of EV for large shocks.

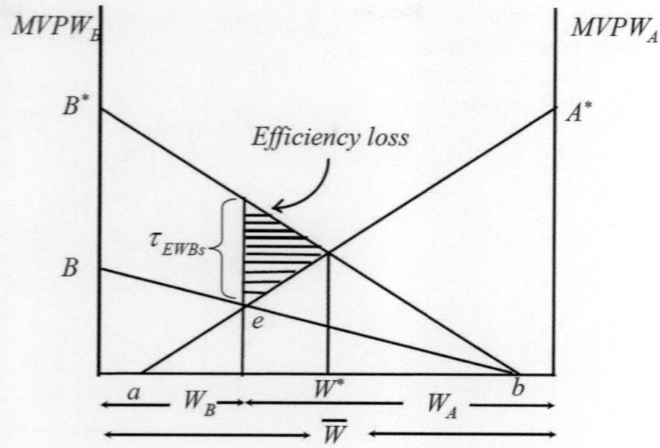

Fig. 2.1 Welfare cost of an implicit subsidy on water used in agriculture (A)

the initial equilibrium denoted by point *e*. Without the tax/subsidy, the equilibrium allocation would be *W**. The loss in economic efficiency to this economy from this distorted use of water is measured by the shaded area in Fig. 2.1.

Several CGE studies have sought to quantify this loss in efficiency. For example, in a recent study of water and economic growth in South Africa, Hassan and Thurlow (2011) estimate that the benefits of reallocation of water within the agricultural sector and across water board regions within the country would amount to a recurring economic gain equal to 4.5% of agricultural GDP. This effect is captured in Eq. (2.1) by the term: $\tau_{Eijs} PFE_{ijs} \mathrm{d}QFE_{ijs}$, wherein the implicit subsidy of endowment i to a specific crop (j) in region s is captured by the term: ($\tau_{Eijs} PFE_{ijs} < 0$) due to the presence of a subsidy (negative tax), so that when the amount of water allocated to this use falls, $\mathrm{d}QFE_{ijs} < 0$, there is a welfare gain, as anticipated by Fig. 2.1. The same concept applies to the other terms in Eq. (2.1), which refer to output subsidies/taxes, as well as consumption subsidies/taxes. Intermediate inputs have been excluded from Eq. (2.1) for the sake of simplicity, but can also play a key role. In short, the larger the initial distortion ($\tau_{Eijs} PFE_{ijs}$), and the larger the reallocation of water ($\mathrm{d}QFE_{ijs}$), the greater the potential gain from a reallocation.

To the extent that increasing water scarcity induces such reallocations of water amongst competing uses, then some of this area can be recouped as an efficiency gain, thereby offsetting some of the loss associated with the water reduction. Another way of thinking about this is to consider a situation in which all of the water shortage is absorbed by sector A.

Since the MVP of water in A is less than the average valuation of water in the economy, part of the loss calculated in the first term of Eq. (2.1) will be made up through the reallocation effect captured in the second term of (2.1). This outcome is consistent with the view of some water analysts that irrigated agriculture is the residual claimant on regional water resources, and it is this sector which will suffer most of the reductions if and when water use is curtailed.

Even if there is not an explicit decision to reallocate water between the two sectors, the presence of this factor market distortion can give rise to unanticipated efficiency

Fig. 2.2 Welfare implications of a rise in world food prices facing an agricultural exporter

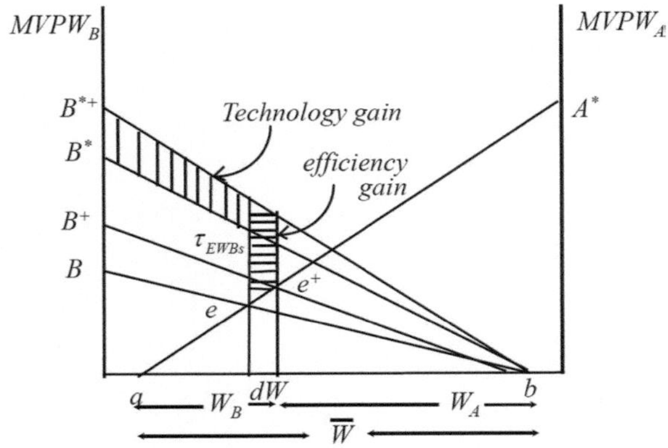

changes. This is provided the allocation of water between the two sectors is subject to adjustment at the margin, as would be the case if this is achieved via a subsidy, a tax, or a quota which is periodically re-evaluated. Consider, for example, the case in which there is an improvement in nonfarm technology over time, such that the marginal value product of water schedule associated with production in sector B rises from B^*b to $B^{*+}b$. Then region s will firstly benefit from improved technology (the shaded technology gain area in Fig. 2.2), and it will also gain from the induced reallocation of water from sector A to sector B (efficiency gain area in Fig. 2.2). This kind of interplay between economic growth and water scarcity on the one hand, and pre-existing distortions on the other hand, is evidenced in the next three terms in Eq. (2.1). The presence of output subsidies or taxes, consumption taxes on domestic and imported goods, as well as tariffs on imports, will likely interact with the quantity changes induced by water scarcity and may give rise to significant welfare effects at the regional level.

Of course, when the inter-sectoral allocation of water is controlled by quotas, and these quotas are not adjusted over time, then the efficiency change component of Fig. 2.2 will not materialize. However, if the pressure caused by the economic growth in sector B eventually results in a reallocation of quotas, then this principle is applicable. There is an additional efficiency gain, relative to baseline welfare in the economy and it is larger, the greater the initial distortion in the water market τ_{EWBs}, and the larger the reallocation of water from sector A to sector B (dW).

The final two terms in Eq. (2.1) refer to the terms of trade effects alluded to above. In practice, most countries engage in two-way trade, such that they are both exporters and importers of most product categories. This makes the terms of trade calculations more complex. It is no longer sufficient to simply focus on a region's net exports. We now must consider the effect of bilateral changes in export prices ($PFOB_{irs}$) and import prices ($PCIF_{irs}$) for all goods traded with all partner regions. Indeed, in an investigation of the trade impacts of projected water scarcity, Liu et al. (2014) find that trading countries are differentially affected depending on how intensively they trade with the economies affected by water scarcity.

From the point of view of policy analysis, the key point of the preceding technical discussion is that water scarcity and economic growth can interact in a variety of ways. Most decision makers think first and foremost about the direct effect: The economy now has fewer resources to work with, therefore growth is expected to slow down, with the extent of this lost welfare depending on the marginal economic value of water to the economy and the size of the shortfall. However, there are many other potential avenues through which water scarcity can affect the economy. By raising the cost of production for water intensive goods and services, water scarcity sends a signal to reduce the output of these sectors. If this releases water to higher value uses, such that most of the adjustment in use occurs in low value (subsidized) sectors, then the losses may not be as great as initially thought. The size of such offsetting gains increases with the size of the initial disparity in effective prices paid for water and the quantity of reallocation which occurs. In her review of the empirical literature on water use and allocation, Olmstead (2013) cites intersectoral price differentials between agriculture and urban uses in the United States where the latter can be paying *as much as 100 times* as much as the former sector pays for water. This suggests that such reallocation benefits could be very large indeed – especially if the water crisis led this distortion to be reduced.[2]

In a global economy, water scarcity also affects the price of traded goods and services. For countries which rely heavily on water-intensive imports, future water scarcity can result in terms of trade losses – or gains in the case of net exporters of these products. If these activities are themselves recipients of other sorts of taxes and subsidies, there is potential for additional 'second best' effects. For example, in their analysis of the impact of future water scarcity on global trade and economic welfare, Liu et al. (2014) find that global welfare losses in their 2030 scenario are exacerbated by the increase, relative to baseline, in subsidized agricultural production in the EU and the US. These two regions experienced relatively less long run scarcity in water in aggregate according to the long run projections (Rosegrant et al. 2013).

2.1.3 Implications for the Modelling of Water Scarcity and Growth

This overview of some of the basic principles behind water scarcity and economic growth provides some important guidelines as we review the literature pertaining to specific sectors of the economy. The first point is that we must factor in the costs of increasing water withdrawals, and this cost function should be convex, such that marginal costs of extraction rise with larger withdrawals. There is empirical support for this characterization of costs; if they are not present in the model, one risks greatly overstating the benefits from increasing withdrawals. A second point relates

[2]Of course what matters here is the marginal price of water being paid. In urban areas, a significant part of the water charge may be due to fixed connection fees. In this case there may be a significant difference between the marginal and average prices of water.

to the importance of understanding the extent to which capital, and other inputs, can be substituted for water in response to scarcity. This could come in the form of more efficient equipment, improved recycling of water, or enhanced extraction. Underestimating this potential for capital-water substitution could lead to overstating the costs of water scarcity to the economy.

The welfare decomposition discussed above underscores the critical importance of accurately estimating the marginal value product of water in different uses and understanding the mechanism by which water shortfalls are allocated across uses. While quotas are a common vehicle for allocating water to different uses, these are themselves likely to change in response to long run market conditions and these changes must be part of any long run analysis as they will have important implications for economic efficiency and hence economic growth.

2.2 Modelling Irrigation Water in a Global CGE Framework

2.2.1 Competition for Water

The major interest of the CGE models concerned with water centers on the role of water as a primary production factor, although a few exceptions examine water as a tradable good or intermediate input. Water can be incorporated into a CGE model either implicitly or explicitly. If implicitly, the model does not represent water directly, but allocates irrigated land which itself embodies water. Modelling water explicitly is more difficult due to the lack of information on water use, its valuation and market pricing. Different strategies have been pursued to circumvent this challenge. The global FARM model (Darwin et al. 1995) explicitly assigns to water a price and a value based on estimations undertaken in the US. Berrittella et al. (2007) use the degree to which water supply falls short of demand to signal the presence of economic rents associated with water. Recent studies argue that an increase in yield on irrigated lands must be achieved to pay for the returns to water. Thus, the shadow price corresponding to the water constraint should be equated to the gains driven by the yield gap between irrigated and non-irrigated production (EPPA-IRC model by Baker (2011); GTAP-W by Calzadilla et al. (2011); GTAP-BIO-W by Taheripour et al. (2013); GTAP-Water by Haqiqi et al. (2016)).

2.2.2 Structure of the Production Function

In the tradition of CGE models, irrigated agricultural production is generally represented by a nested CES production function. An early global CGE model with water developed by Berrittella et al. (2007) has a Leontief production function, which is a special case of the CES function with zero elasticity of substitution. Water and other inputs enter into the final output production function in fixed proportion. Sectoral demand for water responds to a water rent, which is triggered once water supply falls short of demand. The rent increases proportionally with the water gap.

Calzadilla et al. (2011) revised the Berrittella et al. (2007) model and adopted a more flexible three-level CES production function. The bottom level combines land and irrigation water to produce an irrigable land-water composite. In the middle, this composite is further combined with rainfed land and other primary inputs to produce a value-added input composite. The final output is produced from aggregated value-added and intermediates at the top level of the nesting. Such a multi-level nesting structure permits differing degrees of substitutability between inputs at each layer. This model creates two pathways for water to be reallocated between competing uses as water becomes scarce. One is to substitute other inputs for water (a direct effect on water demand); the other is to reduce the demand for water-intensive products (an indirect effect on water demand). However, this pioneering work has significant limitations. The fundamental weakness is that rainfed and irrigated production are treated as part of the same aggregate, national production function. So it is not possible to shut down irrigation in one region in favour of rainfed agriculture, or expanding irrigation in another region. As will be discussed below, based on experience with the TERM-H2O model in Australia, it appears to be the case that such a specification results in an understatement of the potential for reductions in irrigated water use in the face of prolonged drought.

A recently developed model, GTAP-BIO-W, was introduced by Taheripour et al. (2013). It retains the advantages of Berrittella et al. (2007), i.e. the multi-level CES structure and irrigated/rainfed crop production, but overcomes some of its shortcomings. The most marked difference is that GTAP-BIO-W permits competition for resources to take place at two different levels—competition for water within river basins and competition for land within agro-ecological zones (AEZs) between differing rainfed and irrigated sectors. This design significantly improves the adaptability of the model. Since the irrigated and rainfed production functions operate independently from one another, irrigated crop production can be completely removed from a certain part of the country if water supply for irrigation falls short. Moreover, in GTAP-BIO-W, intersections between different river-basins and AEZs are featured by different technologies (i.e. production functions) that reflects water availability, growing conditions and soil quality peculiar to that area (Fig. 2.3).

The shadow value of water in the GTAP-BIO-W model is obtained from the difference in returns to land in the irrigated sector as opposed to its rainfed counterpart within the same AEZ and water basin. This approach is founded upon the assumption that both sectors have the same basic cost structure for non-land inputs. This follows from the assumption of identical input-output ratios (e.g., identical kg. nitrogen fertilizer/ton of crop output) as well as identical non-land input prices. Subtracting the aggregated non-land input cost share (which is therefore equal for rainfed and irrigation agriculture) from the total (which equals one) yields the cost share of land and water in both sectors. Further, because output per unit of land (tons of crop/hectare) is higher when irrigation is applied, land rent per unit of irrigated land

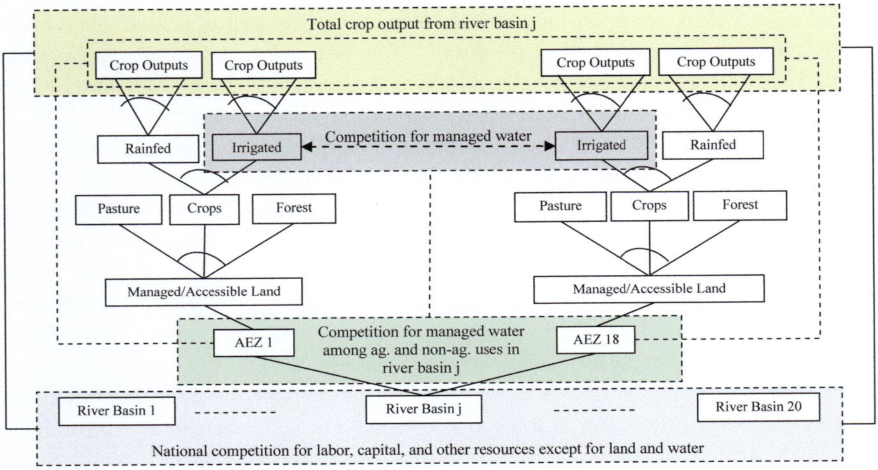

Fig. 2.3 Competition for land and water in the GTAP-BIO-W model. *Source* Taheripour et al. (2013)

is also higher, given the same cost share of land and water. The 'bonus' rent in the GTAP-BIO-W data base is then attributed to the contribution of irrigation water to total production.[3]

These models are among the few global CGE models with water broken out. Incorporating water into a CGE framework has been more extensively explored in studies undertaken at the country and regional level (Darwin et al. 1995; Decaluwe et al. 1999; Dixon et al. 2011; Gómez et al. 2004; Peterson et al. 2005; van Heerden et al. 2008). Most of them still favor the multi-level CES structures. Reviews of these models and the related applications can be found in Calzadilla et al. (2011); Dudu and Chumi (2008) and Ponce et al. (2012). Here, it is useful to draw on one of these single-region studies, namely that of Dixon et al. (2011), who use the TERM-H2O CGE model to examine the economy-wide impacts of a water buy-back policy in the Murray-Darling River Basin of Australia.

Figure 2.4 displays the key portion of a representative agricultural production in the TERM-H2O model. Irrigated activities combine water with un-watered, irrigable land in fixed proportions to produce irrigated land. Irrigated land substitutes for dry land (there is little of this in the irrigated sector) and un-watered, irrigable land, which can also be used in production, albeit with much lower productivity. When a water shortage arises, due to a temporary drought, or a permanent buy-back scheme by which the government reduces the amount of water available for irrigation, then less irrigable land can be irrigated. This increases the availability of un-watered, irrigable land, which is now likely to move into a dry land sector given the prospect of rising commodity prices. The dry land sector uses no irrigated land (by definition) but it

[3]See the derivation in (Hertel et al. 2009).

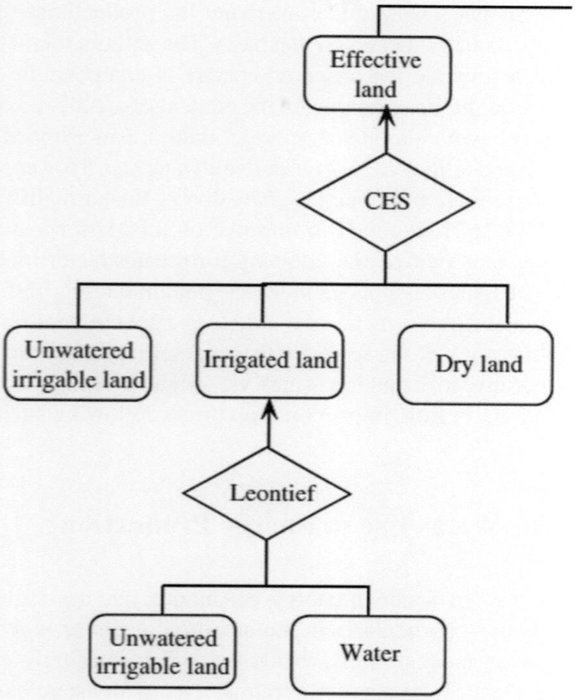

Fig. 2.4 Agricultural land composite in TERM-H2O. *Source* Dixon et al. (2011)

can substitute the un-watered irrigable land for dry land, thereby expanding area the area of rainfed production.

It should be noted that the presence of un-watered, irrigable land in the irrigation production function does not mean that the authors of TERM-H2O can avoid the simultaneous specification of rainfed production sectors. Indeed, as with GTAP-BIO-W, the authors carry along in the model both rainfed and irrigated sectors for all agricultural commodities produced in each of the model regions where irrigation is present. However, the separate identification of currently irrigated, and potentially irrigable land, gives their model a level of nuance which does not exist in GTAP-BIO-W. In the latter model, there are only two types of land, with the land mobility parameter doing double-duty as a regulator of land movement as well as irrigation preparation. In practice, obtaining separate estimates of land equipped for irrigation and actively irrigated land, at global scale, is not currently possible. But doing so would enrich the analysis of global water scarcity.

By separating rainfed and irrigated sectors, the authors of TERM-H2O and GTAP-BIO-W increase the dimensions of their models considerably (2 × the number of irrigated sectors). This raises a natural question: What is gained by this added dimensionality? Some valuable insights are offered in the context of the TERM-H2O model (Wittwer 2012). Early versions of that model did not distinguish irrigated and dry land sectors. In the context of water availability reductions due to drought, separate analyses were undertaken and the productivity of the agricultural sectors was exogenously shocked on that basis. The authors found that the ensuing model greatly underestimated the observed change in composition of farm output in response to the drought. For example, in the context of a 2002–03 drought in the Murray-Darling Basin in which total water usage in the Basin dropped by 29%, the model predicted modest declines in rice output and water use. However, in practice, water usage for rice production dropped by 70% during this drought! The subsequent modifications to TERM-H2O aimed to improve on this performance by separating irrigated and dry land activities and allowing for greater factor mobility between them (Wittwer 2012). Other insights included the potential terms of trade gains to farmers, as sellers of water rights, when water sold to the government for environmental purposes bids up the price of the resource they are selling. They are also able to separate changes in returns to irrigation water vs. irrigable land, which offset one another thereby resulting in little impact on crop prices following such sales.

2.3 Water Use in Energy Production

The role of water in energy production has been attracting considerable attention recently – particularly in the emerging economies where energy demand has been growing most strongly. While water is increasingly used in the extraction of fossil fuels (e.g., fracking of natural gas), its main use remains in electricity production.

2.3.1 Water Use for Hydropower Production

The revelation that hydropower production carries a significant water footprint flows from the fact that hydro-electric dams hold water in a reservoir from which water subsequently evaporates while it is waiting to power the generators. Mekonnen and Hoekstra (2011) find an average water 'footprint' of hydro power generation of 68 cubic meters/GJ of power generated. However, this varies greatly depending on the size of the reservoir surface, relative to installed capacity. The smallest footprint is 0.3 cubic meters/GJ in Colombia, and the largest is 846 cubic meters/GJ in Ghana. Overall, the 35 plants studied (accounting for just 8% of global hydropower capacity), have a water footprint of 90 Giga-cubic meters/year, which is a very significant use of water. This is also a use which is unlikely to adjust in response to water scarcity—short of dismantling the reservoir. This adds to the already significant irreversibilities associated with construction of large hydro facilities.

Implications for CGE modelling: Expansion of hydro-power in the future will depend on a host of factors, including: suitable sites for such plants, the cost of fossil fuels and conventional power generated based on fossil fuels, as well as environmental considerations which are increasingly salient. Construction of a new hydro-power facility is inherently a 'lumpy' decision, with large elements of irreversibility. For this reason, in CGE analysis hydro-power is likely to be scenario-driven – as opposed to a continuously variable response to future developments in energy prices. From the point of view of water scarcity and economic growth, it is critical to factor in the potential losses from evaporation under future scenarios involving significant additions to (or subtractions from) hydro-power. Such estimates could be computed endogenously using the Penman-Monteith equation, as in Mekonnen and Hoekstra (2012), provided the underlying biophysical data are available. As an alternative, one could map any new facilities to the set of 35 facilities for which such calculations have already been done in order to get an idea of how this will likely affect water availability downstream from the facility.

2.3.2 *Water Consumption by Conventional Power Plants*

Rapid economic growth in many developing countries is contributing to significant growth in the demand for electricity, so this source of water demand is projected to greatly increase in the coming decades. The choice of technique for new power plants offers great scope for improved water use efficiency (once the plant is built, there is little scope for modification). The IGES (Bhattacharya and Bijon 2013) estimates that simply by switching from the open loop wet cooling systems favoured historically to a more efficient, closed loop system, India could reduce its requirements from 20% to just 7.5% of total utilizable water. These potential water savings are a concrete manifestation of the potential for capital-water substitution of the sort discussed in the introductory section of this chapter. By investing more in the new technology, very significant water savings can be achieved. However, these large gains are only economical if they are designed in when the plant is built.

Implications for CGE modelling: In order to capture the key aspects of water use in electric power generation described above, there are two critical requirements in a CGE model. The first is a distinction between old and new capital – often described as a 'putty-clay' model. In the context of an existing power plant's infrastructure, there is little that can be done to conserve water use. However, for new power plants, modest increases in capital expenditures can result in significant water savings, as one moves from open to closed loop systems and from wet to dry cooling technologies. Given the cost shares of capital and water in the initial data base, one can choose the elasticity of substitution between capital and water to mirror this abatement cost curve. (This is most readily achieved by simulating the CES function in 'partial equilibrium' mode, i.e. as a cost minimization problem, with output level and input prices fixed. See Golub et al. (2009) for examples in the context of GHG mitigation costs.) In this way, as water becomes scarce, the model will capture endogenous switching to greater water use efficiency.

2.3.3 *Water for Biofuels*

The most important contribution of water to the production of biofuels is through the production of the feedstock. Gerbens-Leenes et al. (2009) explore the water footprint of biofuels for transportation under the IEA's Alternative Policy Scenario. Based on this IEA scenario, the authors find that biofuels could boost their blue water footprint from 0.5% of available blue water worldwide to 5.5% of available water by 2030. This tenfold increase is quite striking and reflects the heavy utilization of irrigation assumed by the authors.

Implications for CGE modelling: Unlike the study by Geerbens-Leenes et al. (2009), global CGE models offer the possibility to endogenize production, consumption and trade in biofuels. If future oil prices remain flat, and biofuel mandates are reduced, then the amount of feedstock required for biofuels in 2030 may be very modest. This, in turn, will reduce the blue water footprint of biofuels. Similarly, by endogenizing the choice of technique for production of the feedstocks, a CGE model can allow the extent of irrigation to be determined by the model as a function of future water scarcity, as well as competition for land with food crops. Therefore, to capture the water footprint of biofuels, the main requirement is to do a good job modelling crop production. This includes the irrigation choice, as well as the competition between biofuels and petroleum in the transportation sector. The latter will ultimately determine the penetration of biofuels in this sector.

2.4 Residential Demand for Water

When compared to other sectors, the residential demand for water in industrialized countries is relatively well understood. The literature suggests that the price elasticity of demand for household water consumption is likely significant – as households appear to have considerable scope for achieving water conservation, given sufficient economic incentive. The average of these estimates is −0.38, with a most likely range between −0.35 and −0.45 in the short run, and −0.55 to −0.75 after accounting for long run investments. While residential water demand in the industrialized countries is reasonably well understood, the same cannot be said of household water demand in the developing countries. This is due to the fact that, in low income countries, households have access to multiple sources of water, and it is unclear what price the household actually pays for the water (much of the cost may be attributable to labour) and the quantity consumed is also much more difficult to measure than in the industrialized countries. While estimation of water demand in the context of developing countries is extremely challenging, Nauges and Whittington (2012) conclude their survey by suggesting that "most estimates of own-price elasticity of water from private connections are in the ranges from −0.3 to −0.6, close to what is usually reported for industrialized countries" (p.264). Those authors conclude their survey by highlighting the need to understand the potential role of dual use networks – one network for drinking water and one for low-quality uses.

Implications for CGE modelling: In principle, it is relatively straightforward to incorporate existing estimates of the price elasticity of the residential demand for water into a CGE model. To do so, one needs a functional form with sufficient flexibility to accommodate this information. The Constant Difference in Elasticities expenditure function is one such option (Hanoch 1975). In this case one would like to calibrate the model to match the long run price elasticity of demand if the goal was to undertake long run growth analysis. The CDE functional form has been successfully employed in the GTAP model over the past two decades (Hertel 1997).

As an alternative, or perhaps in addition to the specification of a household consumption response to water prices, it could be useful to treat water consumption as part of a household production function. In this case, residential consumers would invest in household appliances – with attendant implications for energy and water use efficiency, and then consume the services supplied by this capital stock. In the context of a putty-clay model, such a specification would allow for incorporation of the distinction between short and long run demand responses to water price changes highlighted by Griffin. In the short run there would be limited substitutability between water and other consumption items, but in the long run appliance choice (household capital stock replacement) would have a more profound impact on water conservation.

Another important lesson from this review is the potential value of disaggregating households in developing countries. As Nauges and Whittington (2012) point out, rural and urban households have very different water consumption requirements and the supply of that water has quite different characteristics. Therefore, as demographic change occurs, and the mix of rural and urban households evolves, so too, will the characteristics of aggregate residential demand for water. Breaking out the third group (slum dwellers) would likely be more challenging. Another possibility would be to have a separate module outside of the CGE model which can be used to develop long term projections of household water use and provide an overall summary measure of price response for use in the CGE model. Once the model solution is obtained, one could return to the water consumption module to deduce the impact on rural and urban households.

2.5 Commercial and Industrial Demand for Water

2.5.1 Commercial Water Demand

As noted by Griffin (2006), industrial and commercial water demand is much less intensively studied than residential demand. In the case of commercial demand, while these establishments generally obtain their water from a public utility, the heterogeneity of operations makes estimation of demand difficult in this case. By way of example, Kim and McCuen (1979) highlight the importance of gross store area, length of display windows and drinking fountains in predicting commercial water use. This is a level of detail that is not available to global CGE modellers. Furthermore, even this detailed study did not include estimates of price response. Yet, the rapid growth of the service sector means that it will become increasingly important in overall water use and so its economic behaviour in the face of water scarcity will become increasingly significant. Lacking further information, it seems most reasonable to adapt price elasticities of demand from residential use, as many of the same appliances will be important in commercial use.

2.5.2 Industrial Water Demand

In the case of industrial demand, the challenge lies in the fact that factories are often 'self-supplied', investing in their own infrastructure to access surface or ground water for use in their production processes. Therefore, water use is not monitored, and the relevant price for the water consumed is unobserved. In addition, much of the industrial demand is used for cooling purposes, so that the water flows through the facility and is potentially available for other, downstream uses. Thus it is common in global simulation models to omit the price responsiveness of industrial water demand altogether, as in the IMPACT-WATER model (Rosegrant et al. 2008).

When it comes to consumptive uses, many different production processes rely on some form of steam. Masanet and Walker (2013) investigate in detail the opportunities for reducing steam requirements – which has the great advantage of also reducing energy requirements (since energy is required to convert water to steam)—in the major steam-using sectors of the manufacturing economy: chemicals, paper, petroleum refining and food production. They find ample opportunities for harvesting 'low-hanging fruit' and significantly reducing energy and water usage in these sectors through more efficient use of steam in manufacturing.

What these more efficient engineering processes imply for price responsiveness of the sector is a more challenging question. The approach to demand analysis in the case of industrial use is often through mathematical programming of the industrial operation (Griffin 2006). Such studies may or may not report the implied price elasticity of demand. Griffin (2006) has compiled a table for industrial and commercial price elasticities of demand, based on the work of Renzetti (2002). For purposes of CGE modelling, the most appealing estimates are those undertaken at the two-digit SIC level which range from −0.15 to −0.59 (see also Renzetti (1992)).

Implications for CGE modelling: It is relatively straightforward to 'nest' water in a CES production function in order to elicit a given, output-constant demand response to water prices. A key question is: how should this nesting be done? In many cases conservation of industrial water also results in conservation of energy, suggesting that for some industries these two inputs might be bundled together. And much like energy conservation, improved water efficiency is typically obtained through capital investments. Therefore, nesting water with capital, much as is currently done in energy-oriented CGE models makes sense. Furthermore, allowing some differentiation between short run (clay) and long run (putty) responses would seem appropriate.

2.6 Environmental Demands

It is widely recognized in the water modelling literature that environmental (in stream) uses of water are of critical importance and that outcomes such as biodiversity, depend not only on the total volume of water, but also on the timing and magnitude of high and low flow events (Bell et al. 2014). However, the portion of river flow to be set aside for environmental requirements varies greatly across studies, ranging from 10% in the case of IMPACT-WATER (Rosegrant et al. 2012) to as much as 50% in the IWMI analysis of Smakhtin et al. (2004).

There are many potential contributors to the in-stream value of water flows. Some of these are direct use values, such as contributions to fisheries and wetlands harvest. Ringler and Cai (2006) focus on these two items in their assessment of the value of in-stream flows in the Mekong River Basin in Southeast Asia. Their study highlights the challenges of estimating these values – particularly in the context of developing countries. Their study suggests there are significant costs to further development of the River Basin – for example through further hydropower facilities. Of course, there are also many indirect use values for instream flows, and these are even harder to measure. They include: waste dilution, water storage, flood control and ecological diversity. Brown (1991) reviews studies employing travel cost and contingent valuation methods to value marginal changes in instream flows during times of low flows and finds these to range from about $1/1000 cubic meters to more than $20/1000 cubic meters of flow. These values can exceed the marginal value of irrigation water, suggesting that a reallocation of low flow water from agriculture to the environment would have been beneficial. However, the methods for valuation of these indirect use values are data intensive and subject to significant errors due to the way in which humans respond to such surveys (Huffman 2012), making it difficult to use them on a wide-spread basis.

Implications for CGE Modelling: From a CGE modelling perspective, it would be attractive to move beyond the current approach in which some, apparently arbitrary, percentage of stream flow is set aside as an environmental reserve. A natural way to begin is to incorporate the trade-off between environmental quality and the consumption of other goods and services would be to bring these environmental services into the utility function. One approach, taken by Tsigas et al. (2001) involves specifying a total endowment of environmental quality (in this case, that obtained from maximum stream flow) and then accounting for the use of some of this environmental good (stream flow) for other purposes. What remains is ‘consumed’ by households as an in-stream use of the water. In the initial equilibrium, this in-stream use may not be priced – but rather provided by the government as part of its public services. This specification allows for explicit competition between environmental and commercial uses of water and captures the welfare benefits of improved environmental quality as some of the water is left in-stream for environmental 'consumption' by households.

2.7 Water Supply and Allocation

2.7.1 Reuse of Water

Water is rarely fully consumed in its various uses. Rather, some of the water withdrawn for a particular use is released and becomes a candidate for re-cycling and reuse. Seckler et al. (1998) conclude their report on the future of water supply and demand by suggesting that water reuse may be one of the most important sources of water supply in the coming decades. One of the main barriers to water reuse is pollution, which can build up rapidly as water is recycled. Therefore, the authors emphasize (p. 18) that "pollution control is one of the most basic ways of increasing water supply". In contrast, they find that, at the time of their writing (1998), most international data sets "simply assumed that once water is withdrawn it is lost to further use". This clearly leads to an understatement of water availability.

Luckmann et al. (2014) take account of water reuse in their analysis of water shortages in Israel. They model the supply of seven different types of water separately. This also allows for water to flow out of a sector as both an output (e.g., potable water) and as a by-product which may then be reclaimed for other uses. The intensities of use of these different types of water vary greatly across sectors, and, in many cases, they substitute in use. Thus when freshwater supplies are restricted, there is some potential for shifting to other types of water, albeit at higher cost.

2.7.2 Economy-Wide Water Supply

How should water supply be handled in the context of a global CGE model? A first question is that of regional detail. How much is required? Clearly more detail is better, but at some point the costs of disaggregation exceed the benefits. Chapter 3 of this volume by Liu et al. sheds some useful light on this question. In short, it appears to be quite important that water supplies be modelled at the river basin level, as opposed to the national scale, as might normally be done in a global CGE model.

A second question which arises in the context of CGE modelling of water scarcity and economic growth is whether or not to model the hydrological processes underpinning water supply. It is common in economic models to use the 'tank modelling' approach in which routing is ignored (Bell et al. 2014). In our opinion, this is better left outside the model, with hydrological experts producing estimates of water availability at the river basin level.

Once overall water availability at the river basin level has been determined, there remains the question of how this water gets supplied to users. The municipal supply of water to residential, commercial, and industrial users can be modelled via a production function of the usual sort. The cost shares for this activity should be available from the national input-output table or social accounting matrix. Valuing the water input into this public utility will be a challenge, and will ultimately require an assumption about the price of water in the region. This 'raw water' input may come from either surface water or ground water and its overall availability will be derived from the hydrological model.

A further challenge for CGE models is how to spatially allocate the non-agricultural water uses. Distributing manufacturing and service activities across river basins globally would present a major challenge. Therefore, it is probably easiest to simply require that non-agricultural demands at the river basin scale move in fixed proportion to national activity levels, so these can be viewed as inputs into the single, national production function.

Finally, there is the question of how to model ground water and surface water as distinct source of water supply, given that ground water has a clear buffering value when surface water supplies are uncertain (Diao et al. 2008). In the context of a recursive-dynamic analysis, the supply of surface water to agriculture will depend on pre-existing investments in reservoirs, canals and delivery systems which can be reflected with the presence of a fixed factor in the supply function describing the transformation of raw water into irrigated water. Groundwater supplies are largely dependent on energy costs, which will increase as the depth of groundwater wells increases. The supply of groundwater to agricultural irrigation can be modelled using a production function which combines capital, labour and energy with the ground water endowment. As this endowment is drawn down, the effect on costs can be simulated as non-water, input-using adverse technical change, so that to deliver the same amount of irrigation, more inputs are required. The stock of groundwater will be a function of pumping as well as the rate of recharge. Calculation of the latter will require use of a hydrological model and will therefore need to be done outside the CGE model. Given the differing characteristics of water supplied from surface and ground sources, and in the absence of a more elaborate model of conjunctive use, it is appealing to treat these two sources of water as imperfect substitutes in the irrigation production function, as proposed by Ian Sue-Wing in a recent OECD Working Paper (2013).

2.7.3 Allocation Across Sectors

In the physical science literature, research related to the themes of water availability and allocation aspects usually leans heavily on hydrological modeling (e.g. the LPJmL model (dynamic vegetation, hydrology and crop model) by Gerten et al. (2004), CLIRUN-II (climate runoff model) by Strzepek et al. (2011) and WGHM (WaterGAP Global Hydrology Model) by Döll et al. (2003)) or water management models (e.g. GCWM (global crop water model) by Siebert et al. (2010) and IWSM (integrated global water and food projections model) by Zhu et al. (2013)). The common interest shared by this community is the quantity (and sometimes quality) of water resource. The economic literature, however, provides a different perspective that focuses more on the value of water. In many ways the modelling of water demands and supplies is no more challenging than many of the other demand and supply relationships in the economy. *The real challenge in modelling water has to do with determining its allocation across uses.* As pointed out by Olmstead (2013) water is not typically allocated through markets. This poses a challenge for modellers. For example, in the IMPACT-WATER model (Rosegrant et al. 2012), water is allocated in a sequential fashion, with residential needs satisfied first, followed by industrial and livestock demands. Irrigation needs are treated as a residual claimant on available water. This seems most appropriate in the short- to medium term.

Olmstead (2013) emphasizes the importance of thinking carefully about how the institutions governing water allocations will respond to water scarcity. Where water is allocated on the basis of historical rights and political influence, very large price differentials can emerge across sectors. Given the size of these distortions, one suspects that the impact of reforming the institutions leading to these perverse allocation rules will have a much larger impact than will the anticipated water scarcity itself. On the other hand, water scarcity creates incentives for the reform of water allocation rules and institutions. Olmstead (2013) suggests that "while water prices, on the whole, do not seem to be higher in more arid regions, water marketing is more prevalent in arid regions. In a Coasian sense, the mere existence of the potential gains from trading water creates pressure for trade to occur, so long as the property rights are clearly assigned." In Chile, Australia and the Western US, water trading has evolved in direct response to scarcity (Olmstead 2013).

Among all sectors, agriculture is most likely to relinquish water in the presence of future scarcity, in part due to the fact that agriculture also has the greatest potential to moderate water deficits through supply-enhancing measures (McKinsey & Co 2009). Conservation measures in municipal and domestic use are far more costly and therefore unlikely to come into the optimal mix of policies.

2.7.4 Implications for CGE Modelling

Given the often low opportunity cost of irrigation water and its dominant role in total water withdrawals in the arid regions of the world, it seems logical that agriculture is likely to absorb a large share of future water deficits. This suggests a baseline modelling strategy akin to that of the IMPACT-WATER group in which non-agricultural water use is determined based on projected economic growth and municipal water pricing policies. After running the hydrological model to determine availability at the river basin level, agriculture is be forced to adjust to any ensuing scarcity. This baseline scenario could be contrasted with an efficient allocation scenario in which the price of raw water is equated across sectors.

There are several important issues which a global CGE analysis is likely to have great difficulty in addressing. One of these has to do with return flows. Industrial water is often discharged into streams and rivers. Irrigation water not lost through evapotranspiration may recharge aquifers or add to surface water availability. Proper analysis of these return flows requires considerable spatial detail. And subsequent use of these return flows may require water treatment, in which case one needs to worry about water quality, not just quantity. Studies which ignore the potential for recycling of water, invariably underestimate water supply. Luckmann et al. (2014) make a good start at dealing with reuse by differentiating between seven different types of water, some of which are naturally occurring endowments, some of which are outputs and some of which are by-products of individual sectors. Of course, finding the data to underpin such analysis of reuse at the global scale is likely to be difficult, as pointed out by Seckler et al. (1998).

Another aspect of global water scarcity which is extremely important, but also challenging to come to grips with in a global CGE model, is that of trans-boundary river basins. De Stefano et al. (2010) identify 16 'at risk' river basins which are likely to pose particular problems in the coming decades due to expected increases in hydrological variability and poor institutions governing international flows of water. Ten of these are in Africa—a region already torn by strife. Analyzing these flows and the resulting tensions would be a valuable contribution, but may best be done in a regional model which combines hydrological, climatic and economic dimensions of the problem.

2.8 Conclusions and Research Priorities

The purpose of this introductory chapter has been to survey the literature bearing on various dimensions of water scarcity and economic growth and speak to the implications for CGE modelling of water. Based on the evidence assembled to date Barbier (2004), there appear to have been relatively few instances in which water scarcity has significantly slowed the long term rate of *national* economic growth. However, at local scale, water shortages can have a devastating impact – particularly in the near term, with power outages, retirement of irrigated crop land and unemployment. These localized impacts suggest the need for greater disaggregation than is usually the case in global CGE models. Implementation of a global CGE model with water demand and supply fleshed out at the level of river basins will be a significant undertaking. Chapter 3 of this volume by Liu et al. will show some progress that has been made in this regard. However, there is a great deal of existing data and modelling work outside the CGE community which can form the foundation for such studies (e.g., Rosegrant et al. 2013).

As we look ahead to a world of increasing water scarcity, a key factor will be the scope for society to conserve water: (a) by increasing efficiency in existing uses (e.g. water-efficient appliances), (b) by substituting away from water intensive production activities (e.g., shifting from irrigated to rainfed cropping) and (c) by substituting away from water intensive consumption goods and services (e.g., green lawns). The overall capacity of the economy to substitute increasingly abundant physical and human capital for scarce water, is captured by the elasticity of substitution between these two inputs, σ. If technology and preferences result in a value of $\sigma < 1$, then as water becomes more scarce, the associated economic rents will claim a larger and larger share of GDP. Eventually this could become a significant drag on the economy. In the literature review provided here, it appears that, at the level of individual sectors, this relatively inelastic response to water scarcity is indeed prevalent. However, it remains to be seen what scope there is for the economy to substitute more aggressively away from water intensive activities and consumption goods.

Having flagged the potential for water scarcity to become a brake on economic growth, it should also be pointed out that water use is by no means destined to grow in proportion to population and/or output. Indeed, in reviewing the literature on water demand, we are struck by the ample opportunities for conserving water across the board, including in the electric power sector, the production of industrial steam, residential consumption, and irrigated agriculture. In our opinion, the main reason why such substitution has not been more widespread to date is due to the absence of economic incentives for conservation. In many uses around the world, water remains virtually free. And where a pricing structure does exist, it varies widely across sectors. Indeed, one could argue that the inter-sectoral price differential for water ($100 \times$ in some places) represents one of the most extreme misallocations of a resource in the world economy today.

The presence of this large inter-sectoral distortion heightens the need for general equilibrium analysis. Not only does the CGE approach offer estimates of the direct gains from reducing this distortion, but it also captures how this distortion interacts with other shocks to the economy. Indeed, it is entirely possible that, in a world of increasing water scarcity, water surplus regions could suffer efficiency losses as a result of the interplay between these distortions and international price changes.

The presence of these large factor market distortions is a tribute to the power of the vested interests which control existing water rights. And their resilience over time is evidence of significant political power. However, as water scarcity becomes more severe, it is possible that these vested interests will be overtaken by the broader national interest, and a more economically efficient allocation of water will emerge. In this sense, the looming water crisis could actually become a growth opportunity, since reallocating water to higher value uses could bring significant aggregate benefits.

References

Baker JE (2011) The impact of including water constraints on food production within a CGE framework, Massachusetts Institute of Technology

Barbier EB (2004) Water and economic growth. Econ Rec 80:1–16

Bell A, Zhu T, Xie H, Ringler C (2014) Climate–water interactions—challenges for improved representation in integrated assessment models. Energy Econ 46:510–521

Berrittella M, Hoekstra AY, Rehdanz K, Roson R, Tol RSJ (2007) The economic impact of restricted water supply: a computable general equilibrium analysis. Water Res 41:1799–1813
Bhattacharya A, Bijon M (2013) Water availability for sustainable energy policy: assessing cases in South and South East Asia. Inst Glob Env Strategies
Brown TC (1991) Water for wilderness areas: instream flow needs, protection, and economic value. Rivers 2:311–325
Calzadilla A, Rehdanz K, Tol RSJ (2011) The GTAP-W model: accounting for water use in agriculture (No. 1745). Kiel Working Papers
Darwin R, Tsigas ME, Lewandrowski J, Raneses A (1995) World agriculture and climate change: economic adaptations. USDA, Economic Research Service, USA
Decaluwe B, Patry A, Savard L (1999) When water is no longer heaven sent: comparative pricing analysis in a AGE model. Département d'économique, Université Laval Working Paper 9908
De Stefano L, Edwards P, De Silva L, Wolf AT (2010) Tracking cooperation and conflict in international basins: historic and recent trends. Water Policy 12(6):871–884
Diao X, Dinar A, Roe T, Tsur Y (2008) A general equilibrium analysis of conjunctive ground water and surface water use with an application to Morocco. Agric Econ 38:117–135
Dixon PB, Rimmer MT, Wittwer G (2011) Saving the southern Murray-Darling Basin: the economic effects of a buyback of irrigation water. Econ Rec 87:153–168
Döll P, Kaspar F, Lehner B (2003) A global hydrological model for deriving water availability indicators: model tuning and validation. J Hydrol 270(1–2):105–134
Dudu H, Chumi S (2008) Economics of irrigation water management: a literature survey with focus on partial and general equilibrium models. SSRN Scholarly Paper No. ID 1106504. Rochester, NY
Gerbens-Leenes W, Hoekstra AY, van der Meer TH (2009) The water footprint of bioenergy. Proc Natl Acad Sci 106:10219–10223
Gerten D, Schaphoff S, Haberlandt U, Lucht W, Sitch S (2004) Terrestrial vegetation and water balance—hydrological evaluation of a dynamic global vegetation model. J Hydrol 286(1–4):249–270
Golub A, Hertel TW, Lee H-L, Rose S, Sohngen B (2009) The opportunity cost of land use and the global potential for greenhouse gas mitigation in agriculture and forestry. Resour Energy Econ 31:299–319
Gómez CM, Tirado D, Rey-Maquieira J (2004) Water exchanges versus water works: Insights from a computable general equilibrium model for the Balearic Islands. Water Resour Res 40:W10502
Griffin RC (2006) Water resource economics: the analysis of scarcity, policies, and projects. The MIT Press
Hanoch G (1975) Production and demand models with direct or indirect implicit additivity. Econometrica 43:395–419
Haqiqi I, Taheripour F, Liu J, van der Mensbrugghe D (2016) Introducing Irrigation Water into GTAP Data Base Version 9. J Global Ec Analysis 1:116–155
Hassan R, Thurlow J (2011) Macro–micro feedback links of water management in South Africa: CGE analyses of selected policy regimes. Agric Econ 42:235–247
Hertel TW (1997) Global trade analysis: modeling and applications. Cambridge University Press, New York
Hertel TW, Rose S, Tol RSJ (2009) Land use in computable general equilibrium models. Econ Anal Land Use Glob Clim Change Policy 1
Huff K, Hertel TW (2001) Decomposing welfare changes in GTAP. GTAP Tech. Pap. No 05. http://www.gtap.agecon.purdue.edu/resources/res_display.asp?RecordID=308. Accessed 18 September 2018
Huffman WE (2012) Kahneman's Psychology of Value: the sixth T. W. schultz lecture. Am J Agric Econ 94:285–290
Kim JR, McCuen RH (1979) Factors for predicting commercial water use1. JAWRA J Am Water Resour Assoc 15:1073–1080
Liu J, Hertel TW, Taheripour F, Zhu T, Ringler C (2014) International trade buffers the impact of future irrigation shortfalls. Glob Environ Change 29:22–31

Luckmann J, Grethe H, McDonald S, Orlov A, Siddig K (2014) An integrated economic model of multiple types and uses of water. Water Resour Res 50(5):3875–3892
Masanet E, Walker ME (2013) Energy-water efficiency and U.S. industrial steam. AIChE J 59:2268–2274
McDougall R (2003) A new regional household demand system for GTAP. GTAP Tech. Pap
McKinsey & Co (2009) Charting our water future: economic frameworks to inform decision-making. 2030 Water Resources Group
Mekonnen MM, Hoekstra AY (2011) The water footprint of electricity from hydropower. Hydrol Earth Syst Sci Discuss 8:8355–8372
Mekonnen MM, Hoekstra AY (2012) A Global assessment of the water footprint of farm animal products. Ecosystems 15:401–415
Nauges C, Whittington D (2012) Estimation of water demand in developing countries http://elibrary.worldbank.org/doi/abs/10.1093/wbro/lkp016?journalCode=wbro. Accessed s2 September 2018
OECD (2013) Managing water for green growth
Olmstead SM (2013) Climate change adaptation and water resource management: a review of the literature. Energy Econ 46:500–509
Peterson D, Dwyer G, Appels D, Fry JM (2005) Econ Rec 81:S115–S127
Ponce R, Bosello F, Giupponi C (2012) Integrating water resources into computable general equilibrium models - a survey. Fondazione Eni Enrico Mattei Work, Pap
Renzetti S (1992) Estimating the structure of industrial water demands: the case of canadian manufacturing. Land Econ 68:396–404
Renzetti S (2002) The economics of industrial water use. Edward Elgar Publishing
Ringler C, Cai X (2006) Valuing fisheries and wetlands using integrated economic-hydrologic modeling—Mekong river basin. J Water Resour Plan Manag 132:480–487
Rosegrant MW, Ringler C, Msangi S, Sulser TB, Zhu T, Cline SA (2008) International model for policy analysis of agricultural commodities and trade. http://www.ifpri.org/publication/international-model-policy-analysis-agricultural-commodities-and-trade-impact. Accessed 27 September 2018
Rosegrant MW, Ringler C, Msangi S, Sulser TB, Zhu T, Cline SA (2012) International model for policy analysis of agricultural commodities and trade (IMPACT): model description. International Food Policy Research Institute (IFPRI), Washington, DC
Rosegrant MW, Ringler C, Zhu T, Tokgoz S, Bhandary P (2013) Water and food in the bioeconomy: challenges and opportunities for development. Agric Econ 44:131–150
Seckler D, Amarasinghe, U., Molden D, de Silva R, Barker R (1998) World water demand and supply, 1990 to 2025: scenarios and issues. https://ageconsearch.umn.edu/bitstream/61108/2/REPORT19.PDF. Accessed 27 September 2018
Siebert S, Burke J, Faures JM, Frenken K, Hoogeveen J, Döll P, Portmann FT (2010) Groundwater use for irrigation–a global inventory. Hydrol Earth Syst Sci Discuss 7:3977–4021
Smakhtin VU, Revenga C, Döll P (2004) Taking into account environmental water requirements in global-scale water resources assessments. https://core.ac.uk/download/pdf/6405183.pdf. Accessed 27 September 2018
Strzepek K, McCluskey A, Boehlert B, Jacobsen M, Fant IV (2011) Climate variability and change: a basin scale indicator approach to understanding the risk to water resources development and management
Taheripour F, Hertel T, Liu J (2013) Introducing water by river basin into the GTAP-BIO model: GTAP-BIO-W. GTAP Work. Pap. No 77. URL http://www.gtap.agecon.purdue.edu/resources/res_display.asp?RecordID=4304. Accessed 27 September 2018
Tsigas ME, Gray D, Hertel TW, Krissoff B (2001) Environmental consequences of trade liberalization in the western hemisphere. In: The sustainability of long-term growth: socioeconomic and ecological perspectives. Edward Elgar Publishing, Cheltenham
Van Heerden JH, Blignaut J, Horridge M (2008) Integrated water and economic modelling of the impacts of water market instruments on the South African economy. Ecol Econ 66:105–116

Wittwer G (ed) (2012) Economic modeling of water: the Australian CGE experience. Springer, Dordrecht, New York

Zhu T, Ringler C, Iqbal MM, Sulser TB, Goheer MA (2013) Climate change impacts and adaptation options for water and food in Pakistan: scenario analysis using an integrated global water and food projections model. Water Int 38(5):651–669

Dr. Tom Hertel is Distinguished Professor of Agricultural Economics at Purdue University, where his research and teaching focus on international trade, food and environmental security. Dr. Hertel is a Fellow, and a Past-President, of the Agricultural and Applied Economics Association (AAEA). He is also the founder and Executive Director of the Global Trade Analysis Project (GTAP). The original project model and database is documented in the book: Global Trade Analysis: Modeling and Applications, edited by Dr. Hertel, and published by Cambridge University Press. He has supervised more than forty Ph.D. students and published more than 120 peer reviewed journal articles, along with several dozen book chapters as well as four books. Professor Hertel is the inaugural recipient of the Purdue University Research and Scholarship Distinction Award. He has also received a number of AAEA awards including: Publication of Enduring Quality, Distinguished Policy Contribution, Outstanding Journal Article and Quality of Communication. He has also been Advisor to two Outstanding AAEA Ph.D. and MS theses.

Dr. Jing Liu is Research Economist at the Department of Agricultural Economics, Purdue University. She works on issues at the interface between humankind and environment. Her research seeks to improve our understanding of the economic impacts of environmental changes and our ability to adapt to those impacts. Dr. Liu specializes in using computational and statistical tools to analyze interactions between food production system, land and water. She has also worked across a spectrum of applied economic topics including economic damage of natural disasters, international trade, as well as health and nutrition.

Chapter 3
Analyzing Future Water Scarcity in Computable General Equilibrium Models

Jing Liu, Tom Hertel and Farzad Taheripour

Abstract Starting with an elaborate global CGE model, we investigate three simplifications: (1) tackling global questions in a national level model; (2) collapsing irrigated and rainfed crop production into a single sector; and (3) removing river basin boundaries within a country. In each case, we compare their performance in predicting the impacts of future irrigation scarcity on international trade, crop output, land use change and welfare, relative to the full scale model. We find that, if the research question has to do with changes in national-scale trade, production and welfare changes, it may be sufficient to ignore the sub-national hydrological boundaries in global economic analysis of water scarcity. However, when decision makers have an interest in the distribution of inputs and outputs within a region, preserving the river basin and sectoral detail in the model brings considerable added value to the analysis.

Keywords Model comparison · Sectoral and spatial aggregation · Water scarcity · CGE modeling

3.1 Motivation

Economists building computable general equilibrium (CGE) models have long endeavored to deal with the conflict between model accuracy and scarce research resources. This is nowhere more evident than when it comes to adding water to a global-scale CGE model (Berrittella et al. 2007; Calzadilla et al. 2010). The most obvious reason is that water supply and demand are locally determined, while commodity markets are global in scope. Many economic activities associated with water are confined to geophysical boundaries such as a watershed and/or Agro-ecological zones (AEZs), which may not be well represented by administrative

J. Liu (✉) · T. Hertel · F. Taheripour
Department of Agricultural Economics, Purdue University, 403 W State Street, West Lafayette, IN 47907-2056, USA
e-mail: liu207@purdue.edu

G. Wittwer (ed.), *Economy-Wide Modeling of Water at Regional and Global Scales*, Advances in Applied General Equilibrium Modeling,
https://doi.org/10.1007/978-981-13-6101-2_3

boundaries—the commonly applied level of aggregation in a typical global CGE model. Moreover, within a diverse geophysical region, water availability may vary dramatically, depending on local water demand relative to supply, and the mobility of water resources (Maupin et al. 2014).

In addition to geographical and spatial matters, the appropriate sectoral aggregation of water use appears to be another important concern. Water is a necessary input for almost all economic activities. The relative economic importance of water, however, differs markedly between alternative uses (Maupin et al. 2014). Unlike labor and capital endowments, as well as commodities, water can be hard to move due to impaired water rights, the lack of infrastructure, and the sheer expense of moving water against gravity (Garrick et al. 2013; Young 1986). As a result, the marginal value product of water can vary greatly, even within the same watershed, depending on local institutions and the sector of use (e.g., agriculture, manufactures, services or households) (Olmstead 2014). All these arguments highlight the challenge of developing a water-focused, CGE model—particularly at global scale.

Consider a global model with r regions, i commodities, and j industries. Identifying k river basins within each region and further differentiating z AEZs (representing different types of land and climate conditions) within each river basin can conceivably increase the number of equations to be solved from $r*i*j$ to $r*i*j*k*z$. For each of these additional dimensions, data are needed to describe the initial baseline equilibrium as well as economic behavior. Issues of data availability and quality quickly arise.

Nevertheless, adding more natural resource detail to a global CGE model can be rewarding and worthwhile, as we will show below. Exactly how this is done depends on which variables are of interest to decision makers. Water scarcity modelling provides an excellent context for exploring this issue because, although the degree of scarcity is locally determined, it can have broader impacts on the regional and global economies. This paper focuses on the competition for irrigation water, which accounts for 70% of global freshwater withdrawal. The interaction of irrigation with other non-agricultural water uses, although interesting, is excluded, as competition for non-irrigation water is an underdeveloped area in the existing global modeling frameworks. Our investigation, guided by the following questions, serves the purpose of comparing and evaluating current modeling practices with respect to the incorporation of water into global CGE models:

How much detail is required in order to accurately assess the economic impacts of water scarcity?
Is the additional detail important to all variables?
Are there some cases in which the added detail is not required?

An improved understanding of these questions brings multiple benefits to the CGE community. One among them is to address the concern that models lacking full-blown disaggregation cannot be useful. We show that, under some circumstances, sacrificing detail costs little in terms of analytical accuracy but can translate into enhanced feasibility and viability of CGE modeling of water scarcity. On the other hand, our investigation reveals what kind/level of detail is indeed important and therefore should be preserved. We show that missing these details could be costly to the analysis and subsequent policy recommendations. Finally, this knowledge helps identify and prioritize future data development needs of integrated assessment modeling.

3.2 Literature Review: Water in Global CGE Models

Unlike a partial equilibrium model that focuses on just a few markets at a time, a general equilibrium model explicitly defines constraints and feedbacks from all economic activities. The simultaneous determination of prices and incomes reflects both the direct and indirect effects of water-related policies. Therefore, a general equilibrium model allows a better grasp of the real world implications, and has the distinct advantage of assessing the welfare changes across the overall economy.

The literature on water-focused CGE models has been proliferating over the past decade (see the reviews by Dinar 2014; Dudu and Chumi 2008; Griffith 2012; Ponce et al. 2012). Subject to the nature of the problem and data availability, many existing water-focused CGE models highlight one country or a sub-national region. They normally contain disaggregated sectoral demand for and/or supply of water, especially within agriculture (Diao and Roe 2003; Kahsay et al. 2015). Some also consider non-agricultural water uses (Hassan and Thurlow 2011; Kahsay et al. 2015; Koopman et al. 2015). The sources of water supply include surface water, ground water, desalinated water and even recycled waste water (Gomez et al. 2004; Luckmann et al. 2014). Unlike the sectoral disaggregation that has become conventional, modeling the geographic aspect of water competition has been carried out in various ways. Depending on the resolution of source data, the sub-national water market can be defined by irrigation catchment (Dixon et al. 2011), by wet/dry region (Decaluwé et al. 1999), or by sub-national agricultural areas (Cakmak et al. 2009). In other studies, it is not uncommon to assume an integrated water market at the national level.

These elaborate single-region models have proven to be effective in studying water-related issues that can be addressed in isolation (e.g. changes in domestic policies) (Dinar 2014). However, in the case where the water problem affects multiple regions and/or inter-regional interaction plays an important role in restoring the equilibrium, a single-region model may not be sufficient. Future water scarcity, the main interest of this paper, provides exactly such an example. For this reason, researchers working on this topic have been actively pursuing multi-region CGE models.

The major problem with undertaking a global analysis of water scarcity is the challenge of mapping from sub-national detail (e.g. river basins and AEZs) to national/international output and factor inputs. There does not exist currently a publicly available data set to support such analysis. This challenge has been overcome in a recently developed global water-focused CGE model GTAP-BIO-W (Taheripour et al. 2013) by drawing on a variety of data sets including rainfed and irrigated agriculture (Portmann et al. 2010; Siebert and Döll 2010), cropland area and yields (Monfreda et al. 2008; Ramankutty et al. 2008) and water use by river basin (Cai and Rosegrant 2002). However, most of these data are provided at inconsistent resolutions, using different source data and definitions of key variables. Considerable pre-processing is required, including down-scaling the data to the grid-cell level where they can be reconciled prior to aggregation to river basins, AEZs and regions.

Given these complexities, the global modeling efforts are often required to make significant compromises. One of the most common situations involves merging irrigated and rain-fed crop sectors into a single sector that uses irrigation water as an input. This is the approach taken by Berrittella et al. (2007) when they construct the GTAP-W model based on the GTAP database. In this model, water use intensity by crop is computed as the ratio of water consumption to crop output at the national/regional level. The drawback of this approach is that it may dilute the direct impact of irrigation shortage on crop production, especially for regions highly reliant on irrigation. Table 5.3 in Chap. 5 of this book provides such an example. In reality, depending on institutional arrangements, irrigation activity can be shut down altogether in the case of extreme water scarcity (Wittwer 2012). However, this is not possible in the model wherein irrigation water is but one input into a larger production function.

Another type of cost is incurred when water is assumed to be mobile across river basins within a country—an assumption which we term the "unified river basin assumption". Calzadilla et al. (2010) adopted this approach in their update of Berrittella et al.'s GTAP-W model. Since the entire region is treated as one integrated watershed, it is unable to reflect the various degrees of water stress caused by local water supply and demand. Given these challenges in building a global CGE model with water detail, one might logically ask whether the global dimension is important enough to make the sacrifices necessary on the water front. Are the global impacts of scarcity sufficiently strong to justify these compromises? In this paper we use the recently developed GTAP-BIO-W model as a benchmark to investigate the consequences of these simplifications.

In practice the economy comprises many sectors, each with many different end uses for water. Water can also be processed and reused in many cases. Therefore, an economy-wide estimate of σ needs to reflect the possibility of such reuse. In addition, it must reflect not only the ability to become more efficient in specific end uses, but also the possibility of eliminating some end uses altogether. The economy-wide estimate of σ must also incorporate the potential to substitute away from products produced by particularly water-intensive technologies. These types of intra- and inter-sector substitution relationships are difficult to capture in a single aggregate economic model. Thus, the goal of multisector Computable General Equilibrium (CGE) models is to capture these effects. By including within the model the potential for technical substitution and innovations at a disaggregated level, as well as the potential to substitute away from water intensive intermediate and final goods, CGE models allow for an accurate assessment of the economy-wide potential for substitution of capital and other inputs for water. This is why CGE models will be a focal point of this survey of water scarcity and economic growth.

3.3 Overview of GTAP-BIO-W

3.3.1 Model Structure

GTAP-BIO-W is a multi-region, multi-sector, comparative static extension of the standard CGE GTAP model. A thorough description of the latter can be found in Hertel (1997) and will not be repeated here. GTAP-BIO-W is formally documented in Taheripour et al. (2013). This section revisits the fundamental modifications which are relevant for the present paper.

GTAP-BIO-W retains a multi-level constant elasticity of substitution (CES) production function for crop sectors but now includes two production functions for each crop/region: rainfed and irrigated crops (Table 3.4). The major extensions include a more elaborate crop production function and the explicit incorporation of water resource constraint. Water and land are treated as complementary inputs for irrigated farming activities as shown in Appendix Fig. 3.5. This figure represents the modified production functions in GTAP-BIO-W to distinguish between irrigated and dry-land farming technologies. Each region has up to 18 AEZs and 20 river basins to represent the various sub-national growing conditions. Water balance is achieved independently within each river basin where the supply of water is exogenous. In the case that a river basin stretches across country/region boundaries, the river basin is split into sub-basins. Crop production can move across river basins and AEZs. Water can move across AEZs of a river basin, but it cannot move across river basins. As

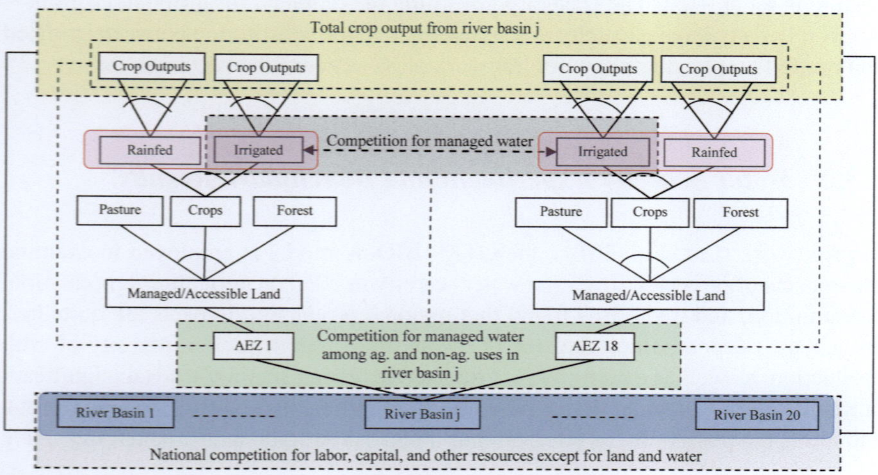

Fig. 3.1 Crop production structure in the GTAP-BIO-W model (Each region contains up to 18 AEZs and 20 river basins. Within a river basin, farming activities from AEZs complete for water. Within each AEZ, irrigated and rainfed sectors compete for land. Within each irrigated and rainfed category, crops compete. Red box indicates where the modification arises for the combined I&R model. Blue box indicates where the modification arises for the unified basin model.)

illustrated by Fig. 3.1, the competition for water and land takes place at two different spatial levels—irrigated crops compete for water within a river basin, while both rainfed and irrigated crops compete for land within an AEZ.

Unlike GTAP-W, which attempts to mimic irrigated production by setting apart the value contribution of water from the original land input, GTAP-BIO-W models irrigated and rainfed activities separately. The two types of production utilize different technologies, yet produce crops which are identical from the consumers' point of view. This design allows for the possibility of completely eliminating irrigated output if water supply for irrigation falls short. Built upon the assumption that irrigated and rainfed activities face the same input-output ratios and non-water-land input prices, the two sectors have the same overall cost structure for non-land-water inputs (more variable inputs are used for each hectare of irrigated land, but output is also higher). Subtracting the non-land-water cost share from the total (which equals one) yields the same cost share of land-water input in both sectors. Importantly, although the cost shares are identical, the land-water returns per hectare are not necessarily the same. Indeed, since output per unit of land with irrigation is typically higher, land return per hectare is higher for the irrigated than the rainfed land. At each river basin, this additional return attributed to irrigation becomes the shadow value of water in the GTAP-BIO-W model.

Sectoral allocation of land follows a CET structure,[1] as shown in Appendix Fig. 3.6. The key parameters regarding land supply and transformation are inherited from previous studies, since parameter estimation is not the central focus of this study. Following Keeney and Hertel (2009) and Hertel et al. (2010), the elasticity of transformation that allocates managed land into cropland, forest and pasture is set at −0.2. The elasticity governing the cropland for different crops is −0.5. A relatively large elasticity −10 is assigned to the substitution between irrigated and rainfed cropland to allow for transformation between the two uses.

3.3.2 *Water Scarcity Experiment and Benchmark Results*

In prior work (Liu et al. 2014), the GTAP-BIO-W model is employed to examine the impacts of projected irrigation water scarcity in 2030 on agricultural production, consumption, and trade. It is found that regional production impacts are quite heterogeneous, depending on the size of the shortfall, the irrigation intensity of crop production, as well as the crop mix. Projected irrigation scarcity leads to significant output declines in China, South Asia, the Middle East and North Africa (MENA), and alters the geography of international trade. These trade adjustments play a key

[1] For example, land supply to crop sectors is determined by crop output, the price of cropland relative to managed land, and the elasticity of transformation $\sigma1$. In GTAP coding, qocropland(i, r) = qo(i, r) − endwslack(i, r) + $\sigma1$ * [pmland(i, r) − pmcropland(i, r)], where i indicates agro-ecological zone (AEZ) and r indicates region.

role in moderating the adverse impact on global prices and consumption—thereby highlighting the importance of examining water scarcity in a global framework.

The present study employs the irrigation scarcity shocks from Liu et al. (2014). The extent of irrigation shortfall is measured by the Irrigation Water Supply Reliability (IWSR) index—the share of potential irrigation demand that is realized through actual consumption, on an annual basis (Rosegrant et al. 2013). The irrigation demand and consumption components were simulated in the IMPACT-WSM model by following a priority-based rule that assumes irrigation water is the residual (see Appendix). Key drivers of water demand and supply include population and economic growth, infrastructure development, and crop evapotranspiration (see Appendix for more information). Climate change is not considered, however, in an effort to avoid additional uncertainty being introduced to the projection.

The water scarcity shock in this CGE model is constructed as the percentage change of the river-basin level IWSR (2030 relative to 2000, Table 3.5). Each region contains up to twenty river basins (see Table 3.3 for basin names). Figure 3.7 shows that in 2030 water will become scarcer in South Asia, North China Plain, and East Africa. It is important to note that water consumption (i.e. evapotranspiration) rather than water withdrawal is used in the IWSR simulation. Inefficient irrigation usually leads to more water withdrawn than actually consumed. To the extent that water consumption and water withdrawals grow at different rates, the IWSR shock may under- or overstate the actual shortage of water for irrigation. Formal discussion about the discrepancy between the two can be found in Falkenmark and Lannerstad (2005).

3.4 Experimental Design: Three Simplifications for Global Water Modelling

3.4.1 Simplification One: Single-Region Analysis (Henceforth the Single-Region Model)

Implementing a single-region specification is accomplished by forming a closure in which production activities in the focus region remain endogenous while the activities in the rest of the world are held constant. We implement multiple slack variable "swaps" to fix income, output and prices in the exogenous regions. Irrigation water supply is shocked in the focus region only. This means that irrigation availability in the other regions/basins remains unchanged—not unlike the approach taken by Wittwer (2012) in the case of Australia. Of course, since production has been held constant in the rest of the world, it does not matter whether the shocks are applied to only the focus region or to the other regions as well.

This single region closure is repeated for China, India, South Asia excluding India (or the rest of South Asia), Brazil, sub-Saharan Africa (SSA) and the MENA region. According to Rosegrant et al. (2013), some water basins in these regions will face significant irrigation shortages by 2030. Appendix Table 3.5 summarizes the shocks in each individual experiment. The rest of the experiment, including baseline data and parameters remain the same as in the benchmark model.

3.4.2 Simplification Two: Collapsing Irrigated and Rainfed Agriculture into a Single Activity (the Combined I&R Model)

This experiment aims to imitate a model which does not differentiate irrigated from rainfed crop sectors. Instead of reversing what has been done for the full GTAP-BIO-W model,[2] i.e. merging irrigated and rainfed sectors in the database, we modify the zero-profit conditions for crop sectors. In the full GTAP-BIO-W model, irrigated and rainfed output are determined separately. Our modification relaxes these individual constraints and replaces them with an equation that synchronizes the percentage change of output in two sectors. To be specific, for every 1% increase in total wheat output, both irrigated and rainfed wheat output increase by 1%. In this way, irrigated and rainfed activities are assured to move together, as would be the case if we had modelled production on one single combined sector. The rest of the model, including base data, parameters and shocks remain unchanged from the full GTAP-BIO-W model.

3.4.3 Simplification Three: Aggregating River Basins to the National Level (the Unified Basin Model)

The key in unifying the sub-national river basins is to equalize the shadow unit value of water within the nation/region. In the benchmark GTAP-BIO-W model, water market is segmented by hydrological boundaries. The value of water is determined separately in each river basin by equating water demand and supply. Competition for water takes place within the basin among irrigated sectors that face the same price for irrigation. In order to eliminate the sub-national river basin boundaries, we modify the market-clearing condition for water such that the value of water at all sub-national basins changes at the same rate. Slack variables are added to break the initial constraint that clears water market at each individual river basin. Additionally, an equation for the entire nation/region is added to determine the shadow unit value of water.

It is important to bear in mind that, the three simplifications listed above are implemented independently. The single-region model maintains the irrigated-rainfed split and the sub-national river basin boundaries. The combined I&R model does not relax

[2]The full model or benchmark model refers to the GTAP-BIO-W model with rainfed-irrigated split and basin boundaries within regions.

the water balance constraint at each river basin. The unified basin model differentiates crop production functions by irrigation condition. The results from implementing each simplification are then compared with the results obtained from the full-blown GTAP-BIO-W model. Four metrics—bilateral trade flows, crop output, harvested area, and welfare are selected to demonstrate the tradeoffs between model complexity and accuracy.

3.5 Analyzing the Consequences of Global Model Simplifications

3.5.1 Impacts of Future Water Scarcity on Trade

We conduct our comparison metric by metric, beginning with the trade flow metric. Figure 3.2 compares the net bilateral trade flows of food (exports minus imports) simulated by different models.[3] We focus on trade flows between the rest of the

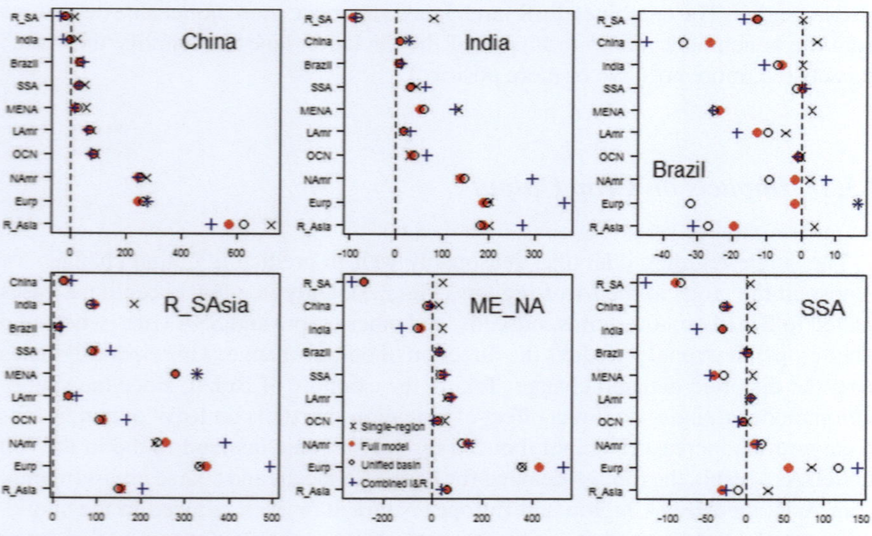

Fig. 3.2 A comparison of changes in net trade flow relative to baseline, under different models (Only food and agricultural products are considered. Included are crops, livestock and processed livestock products, and processed food products. Each plot reflects only the bilateral trade flow to the titled region from the rest of the world. Positive (negative) numbers indicate increase in net exports (imports). Unit is million USD 2001. It is important to note that the range of x-axis varies across the plots.)

[3] All models contain 19 regions. In some tables, we aggregate the original 19 regions into 11 regions (based on weighted summation) for the ease of reporting.

world (RoW) and the six regions involved in the single-region models. First of all, in most cases the single-region model suggests increased net food exports from RoW to the focus region. This is not surprising, because in this model irrigation scarcity does not affect RoW. The comparative disadvantage encourages the focus region to import more food. This is a fundamental shortcoming of the single-region approach if one is interested in using it to analyze trade flows.

Secondly, the single-region model tends to wrongly predict the direction of net trade flow between the focus region and its more water-stressed trade partners. Examples include the flows from the rest of South Asia to China in the single-region model highlighting China, and the flows from the rest of South Asia to India in the single-region model highlighting India. This is because both of the single-region models fail to reflect the fact that rest South Asia faces more severe irrigation scarcity than China and India. By contrast, as for trade flows between the focus regions and their less water-stressed trade partners, the single-region model tends to overstate the net imports increase because the relative degree of irrigation scarcity between regions is exaggerated.

The results of trade flow in the unified basin model exhibit very high similarity to the full model results. The deviation is only relatively larger when trade flow is large, for example, the flow from the rest of South Asia to China, and from Europe to the MENA region. The combined I&R model tends to create more noticeable deviations from the benchmark model results. And the deviation tends to amplify the "true" impacts (i.e. more negative or more positive).

3.5.2 Impacts on Crop Output

The single-region model does reasonably well in predicting output changes for almost all the crops in the focus regions (Fig. 3.3). Only in a few exceptional cases related to Brazil (coarse grains, oilseeds, and other crops) and SSA (rice, wheat and other crops), it wrongly predicts the direction of output change. This error originates from the distorted demand change. Taking the example of Brazil, since the single-region model assumes no direct effect of irrigation shortfalls on RoW output, it fails to capture the increasing output induced by international demand found in the full model. As a result, the foreign demand for Brazil's oilseeds and coarse grains (mainly from Asia, the MENA region and Europe) is muted or even reduced in the single-region model, which is quite understandable because in the context of irrigation shortage Brazil produces these commodities at a higher cost.

Only Brazil and SSA suffer from this error because they are relatively less affected by water scarcity[4] than the other regions that face the same issue. Such a relative advantage can be successfully reflected in the full model but not so well in the single-region model. Instead, it is misrepresented as a disadvantage (over RoW) in the latter.

[4]SSA is not heavily reliant on irrigation. For Brazil, irrigation condition will improve in one of the major agricultural production areas.

Crop output is therefore predicted to increase in one case but decrease in the other. This disagreement, however, does not occur in the more severely affected regions because both models predict that crop output will decline.

In most cases, the unified basin model outperforms the combined I&R model, and provides more accurate predictions on crop output changes. We find that, forcing irrigated and rainfed sectors to move together tends to exaggerate the output loss caused by irrigation shortfalls. This is sensible because rainfed production would have been able to expand and make up for the contracted irrigated output if it were not tied to the latter. Additionally, the magnitude of the error varies across regions and crop sectors. Larger shocks normally lead to more significant deviations from the benchmark model predictions. Except for the rest of South Asia, the output deviation from the benchmark result is less than 2%.[5]

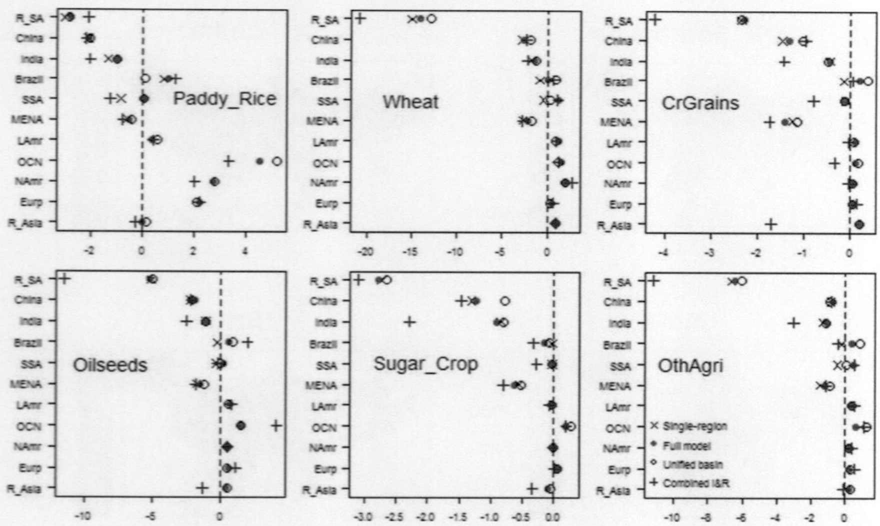

Fig. 3.3 A comparison of changes in crop output relative to baseline, under different models (Results associated with single-region models show only output change in the focus region. Crop output change in non-focus regions is zero by model design and thus omitted from the plot. X-axis refers to percentage point change.)

[5]We also conducted the Welch two-sample t-test to examine whether the results obtained from different models are significantly different. Each sample contains 114 observations (19 regions, 6 crops). We consider two pairs of results: full model versus combined I&R model (t-statistic = 1.63,p-value = 0.11), and full model versus unified basin model (t-statistic = −0.04, p-value = 0.97). In both tests, the crop output changes in each pair are not significantly different from each other at the 5% level, although the combined I&R model produces more similar results (to the full model results) than does its competitor model.

3.5.3 Impacts on Land Use Change

Figure 3.4a, b show how well each simplification preserves the benchmark results with regard to irrigated harvested area. Depending on the severity of basin-specific

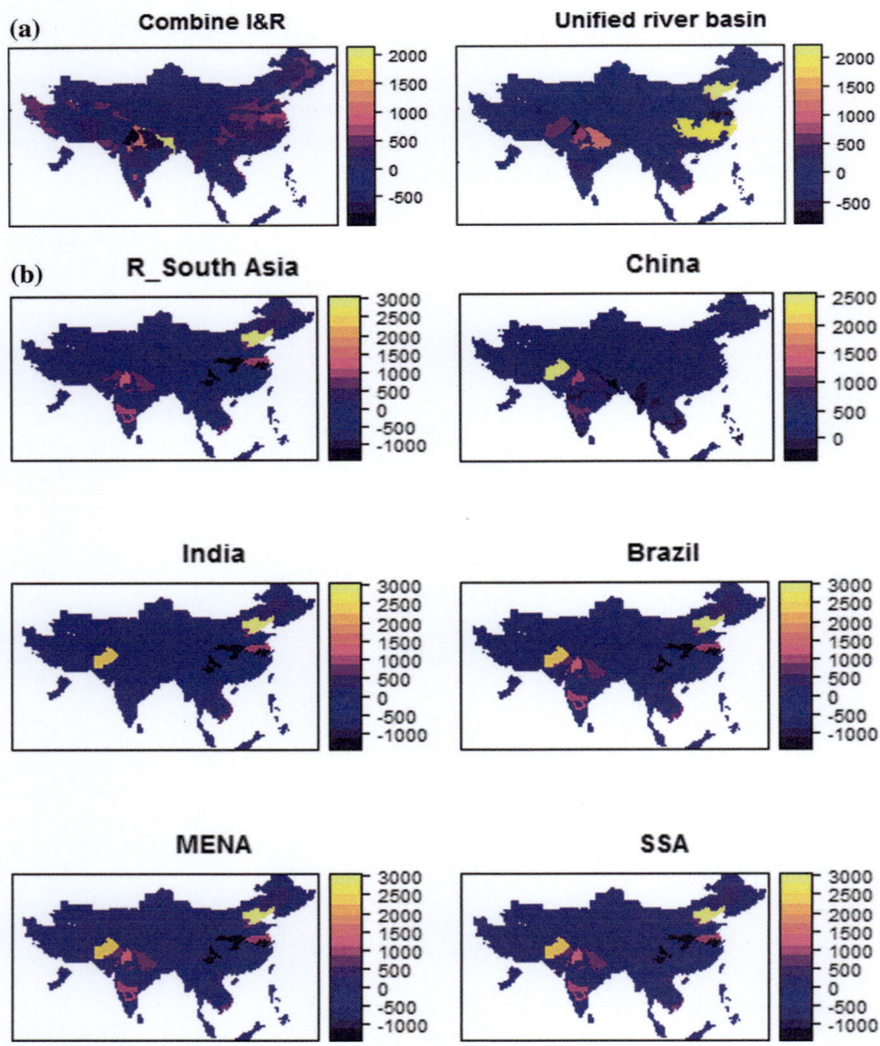

Fig. 3.4 **a** A comparison of the **deviations** of irrigated harvested area from the full model results, combined I&R model and the unified basin model (thousands of hectares). **b** A comparison of the **deviations** of irrigated harvested area generated by the single-region model highlighting each titled region (thousands of hectares)

irrigation scarcity and the water content of crops at each location, the combined I&R model and the unified basin model could either over- or under-estimate the change of irrigated harvested area (Fig. 3.4a). In many cases, they understate the reduction in irrigated area, and the deviations are non-trivial at the sub-national scale.

The divergence in the combined I&R model can be intuitively explained by the fact that the "merged" crop sector approximates a weighted average of irrigated and rainfed sectors.[6] This leads to two potential errors. For the region as a whole, the overall contraction of irrigated area in the combined I&R model becomes smaller; whereas within each river basin, the reallocation effect on irrigated production becomes stronger. That is, irrigation-intensive AEZs lose more irrigated area, while less irrigation-intensive AEZs may harvest more irrigated area.

Similarly, the errors in the unified basin model can be explained by the fact that the "merged" basin approximates a weighted average of all the basins within the region. The reduction of irrigated area suggested by the unified basin model appears to be smaller than in the benchmark model, because river basins are subject to a smaller average irrigation shock in the former. Furthermore, irrigated area is spatially redistributed within the region, as opposed to within a river basin in the combined I&R model. For this reason, extreme biases are more likely to be seen in the unified basin model.

As for the single-region model results, large errors are normally observed outside of the focus region (Figs. 3.4b and 3.8b). These observations conform to the expectation that the single-region model does a better job in capturing the impacts on the focus region but not necessarily the impacts on the rest of the world. Moreover, in cases where irrigation scarcity shock is small (e.g. Brazil, SSA and MENA), the respective models produce very similar results, because they all have very little to say about the other regions.

3.5.4 Macro-Level Variable Changes: Welfare and Trade Balance Comparison Across Models

The benchmark model predicts a world welfare loss due to the reduction in water endowment under the scarcity experiment, as well as the changing efficiency of factor use. The size of the loss, however, is relatively small because the economy adjusts to offset the shock. Table 3.1 (as well as Table 3.6) presents a comparison of the welfare change obtained from each model. The unified basin model clearly outperforms the combined I&R model which significantly exaggerates the welfare loss. Comparing across single-region simplifications, the model normally performs better when the region is subject to severe irrigation scarcity. Table 3.2 compares the change in trade balance of food commodities. Again, the unified basin model provides results that are

[6]Here the weight is the value share of irrigated land θ_r, which is less than one. For example, $qfe(r) = \theta_r qfe(irr, r) + (1 - \theta_r) qfe(rdf, r)$. Contraction of total cropland area in region r will be less pronounced due to the weight or even flipped as the expansion of rainfed cropland becomes strong.

more aligned with the full model. This suggests that, if all one is interested in is the macro-welfare impacts of water scarcity, perhaps a unified basin model is sufficient.

3.6 Summary and Implications for Future Research

Comparing models with differing levels of complexity, we investigate the costs and benefits of simplifications commonly undertaken in the context of global water-related CGE analysis. Our findings are specific to this context. We do not attempt to conclude which simplification is superior to the others. Instead, the choice is determined by the metrics of interest to decision makers.

The single region model typically biases the change of bilateral trade flows. However, it is generally functional in predicting crop output changes in the focus region. Of course it is silent on the production impacts in other regions. Thus, if one is only interested in the impacts on a particular country, and provided the pattern of trade is not paramount to the decision makers, then a single-region model is likely to be sufficient for the analysis of irrigation scarcity.

A multi-region model that merges irrigated and rainfed production results in overstated overall reduction in crop output, but understates the change in irrigated crop output. This echoes what Wittwer (2012) found in the context of the 2002 drought in Australia: failure to differentiate irrigation conditions led to far too small a change in irrigated area. In its comparison with the unified basin model, the combined I&R model in general performs worse in predicting crop output, bilateral trade flows and total welfare change. Dividing agriculture into rainfed and irrigated sub-sectors could enhance analysis related to these aspects of the economy, and this is especially true for the irrigation-stressed regions (e.g. the rest of South Asia).

The unified basin model, however, can be more useful for avoiding extreme errors in predicting the spatial distribution of irrigated harvested area within a region (e.g. China in Fig. 3.4a). This is because the irrigated harvested area can be redistributed to any river basin within the region rather than within the basin. If the decision maker does not care about the location of production and land use, it may be sufficient to ignore the sub-national river basins in global economic analysis of water scarcity. We also find that the unified basin model performs extremely well in predicting national welfare impacts of water scarcity.

On the other hand, high-resolution geophysical data can be valuable if one is interested in the sub-national distribution of changes in production, land use and water. When this is the case, extending a CGE model to reflect the detail becomes important. The major constraint in doing so is data availability. The recently developed GTAP-BIO-W model has made progress in coping with these issues, but also has some limitations. For example, water in GTAP-BIO-W is used by agriculture but not by other non-agricultural sectors. Besides, the role of groundwater is not clearly reflected in the model.

Since it is difficult to incorporate as much detail in global modeling as one might ideally desire, understanding the related tradeoffs provides useful insights into where

Table 3.1 Welfare changes under different models (million USD 2001)[a]

	Full model	Combined I&R model	Unified basin model	Single-region-Brazil	Single-region-China	Single-region-India	Single-region-RSouthAsia	Single-region-MENA	Single-region-SSA
Brazil	128	75	112	2					
China	−995	−2358	−1006		−797				
India	−784	−7184	−789			−751			
RSouthAsia	−1191	−4124	−1119				−1133		
MENA	−481	−719	−395					−142	
SSA	68	−622	51						−19
World	−3726	−15,775	−3497						

[a]For the single-region models, only the impacts to the interested region and the world are reported. The impacts in other regions are not meaningful to be compared with those obtained from the other models, and thus are omitted in the table

Table 3.2 Change in trade balance (million USD 2001) under different models[a]

	Full model	Combined I&R model	Unified basin model	Single-region-Brazil	Single-region-China	Single-region-India	Single-region-RSouthAsia	Single-region-MENA	Single-region-SSA
NAmr	1111	1503	1026	7	376	176	266	148	15
Eurp	998	1701	1031	8	227	152	256	275	52
Brazil	181	238	206	−19	58	16	24	27	4
LAmr	505	684	454	−1	116	60	59	76	11
China	−1069	−1035	−1157	2	−1318	31	42	21	4
India	−478	−1020	−516	0	27	−733	176	33	3
RSouthAsia	−1295	−1918	−1214	0	12	45	−1341	28	2
RAsia	88	−350	149	1	273	98	150	42	10
MENA	−611	−597	−460	1	31	53	121	−757	6
SSA	249	351	137	1	51	41	99	45	−108
OCN	321	444	340	1	129	62	123	55	4

[a]Only food commodities are included. Positive numbers indicate increase in net exports

such applications might be needed. In the long run, improvements in global geospatial data will lessen the costs of constructing water-oriented, global CGE models and we expect that such analyses will become more popular due to their ability to simulate water scarcity at the river basin level—a phenomenon which is expected to become more important, worldwide, in the future.

Appendix

In the IMPACT modeling suite, natural water availability and supply for irrigation are determined with the Global Hydrologic Model (IGHM) and the Water Simulation Model (IWSM), respectively, as illustrated in Fig. 3.9. The IGHM is a semi-distributed hydrological model that simulates evapotranspiration, surface runoff and base flow on 0.5° latitude × 0.5° longitude grid cells over global land surfaces, except for Antarctica. It uses a temperature-index method adapted from NOAA's SNOW-17 model to simulate snowpack accumulation and ablation. Gridded hydrological output is spatially aggregated to the Food Production Units (FPUs), weighted by grid cell areas, for use by the IWSM model.

The IWSM uses monthly runoff and potential evapotranspiration from the IGHM to simulate water management and allocation processes for river basins, using FPUs

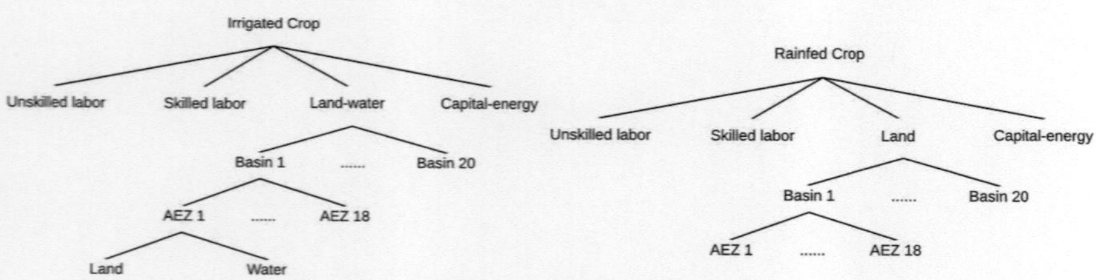

Fig. 3.5 The structure of primary inputs for irrigate and rainfed crop production

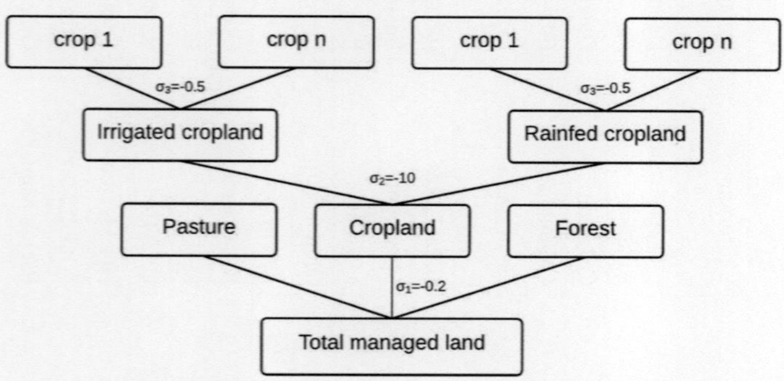

Fig. 3.6 Transformation parameters governing the allocation of managed land in the GTAP-BIO-W model

Table 3.3 AEZs (B1–B20) by GTAP-BIO-W region

	USA	EU27	Brazil	CAN	Japan	CHIHKG	India
B1	Arkansas	Baltic	Amazon	Canada Arctic Atlantic	Japan	Amur	Brahmaputra
B2	California	Britain	N S Amri. Coast	Cnt Canada Slave Basin	Others	Brahmaputra	Brahmari
B3	Canada Arctic Atlantic	Danube	NE Brazil	Columbia	–	Chang Jiang	Cauvery
B4	Colorado	Dnieper	Orinoco	Great Lakes	–	Ganges	Chotanagpui
B5	Columbia	Elbe	Parana	Red Winnipeg	–	Hai He	E Ghats
B6	Great Basin	Iberia E Med	San Francisco	US NE	–	Huai He	Ganges
B7	Great Lakes	Iberia W Atlantic	Toc	MacKenzie	–	Huang He	Godavari
B8	Mississippi	Ireland	Uruguay	Pacific Namer N	–	Indus	India E Coast
B9	Missouri	Italy	Others	Others	–	Langcang Jiang	Indus
B10	Ohio	Loire Bordeaux	–	–	–	Lower Mongolia	Krishna
B11	Red Winnipeg	N Euro Russia	–	–	–	N Korea Peninsula	Langcang Jiang
B12	Rio Grande	Oder	–	–	–	Ob	Luni
B13	SE US	Rhine	–	–	–	SE Asia Coast	Mahi Tapti
B14	US NE	Rhone	–	–	–	Songhua	Sahyada
B15	Upr Mexico	Scandinavia	–	–	–	Yili He	Thai Myan Malay
B16	W Gulf Mex	Seine	–	–	–	Zhu Jiang	Others
B17	Pacific Namer N	Others	–	–	–	Mekong	–
B18	Others	–	–	–	–	Others	–
B19	–	–	–	–	–	–	–

(continued)

Table 3.3 (continued)

	USA	EU27	Brazil	CAN	Japan	CHIHKG	India
B20	–	–	–	–	–	–	–

	C America	S America	E Asia	MLYS & IDN	R. SE Asia	R. S Asia	Russia
B1	Caribbean	Amazon	Amur	Borneo	Borneo	Amudarja	Amur
B2	Cnt Amri.	Chile Coast	N Korea Peninsula	Indonesia E	Langcang Jiang	Brahmaputra	Baltic
B3	Cuba	NE S Amri.	S Korea Peninsula	Indonesia W	Mekong	Ganges	Black Sea
B4	Middle Mexico	NW S Amri.	Lower Mongolia	Papau Oceania	Philippines	Indus	Dnieper
B5	NW S Amri.	Orinoco	Upr Mongolia	Thai Myan Malay	SE Asia Coast	Sri Lanka	Lower Mongolia
B6	Rio Grande	Parana	Others	Others	Thai Myan Malay	Thai Myan Malay	N Euro Russia
B7	Upr Mexico	Peru coastal	–	–	Others	W Asia Iran	Ob
B8	Yucatan	Rio colorado	–	–	–	Others	Scandinavia
B9	Others	Salada Tierra	–	–	–	–	Upr Mongolia
B10	–	Tierra	–	–	–	–	Ural
B11	–	Uruguay	–	–	–	–	Volga
B12	–	Others	–	–	–	–	W Asia Iran
B13	–	–	–	–	–	–	Yenisey
B14	–	–	–	–	–	–	Siberia Other
B15	–	–	–	–	–	–	Others
B16	–	–	–	–	–	–	–
B17	–	–	–	–	–	–	–
B18	–	–	–	–	–	–	–
B19	–	–	–	–	–	–	–

(continued)

Table 3.3 (continued)

	C America	S America	E Asia	MLYS & IDN	R. SE Asia	R. S Asia	Russia
B20	–	–	–	–	–	–	–

	E-Europe-RFSU	R. Europe	M-East-N-Afri,	SSA	Oceania
B1	Amudarja	Rhine	Arabian Peninsula	Cnt Afri. W Coast	Cnt Australia
B2	Amur	Rhone	Black Sea	Congo	E Australia Tasmania
B3	Baltic	Scandinavia	E Med	East Afri. Coast	Murray Australia
B4	Black Sea	Others	Nile	Horn of Afri.	New Zealand
B5	Danube	–	North Afri. Coast	Kalahari	Papau Oceania
B6	Dnieper	–	NorthW Afri. Coastal	Lake Chad Basin	Sahara
B7	E Med	–	Sahara	Limpopo	W Australia
B8	Iberia E Med	–	Tigris Euphrates	Madagascar	Others
B9	Lake Balkhash	–	W Asia Iran	Niger	–
B10	Lower Mongolia	–	Others	Nile	–
B11	Ob	–	–	NorthW Afri.	–
B12	Syrdarja	–	–	Orange	–
B13	Tigris Euphrates	–	–	Sahara	–
B14	Upr Mongolia	–	–	Senegal	–
B15	Ural	–	–	South Afri. Coast	–
B16	Volga	–	–	Southeast Afri. Coast	–
B17	W Asia Iran	–	–	Volta	–
B18	Yenisey	–	–	W Afri. Coastal	–
B19	Yili He	–	–	Zambezi	–
B20	Others	–	–	Others	–

Table 3.4 All sectors covered by the GTAP-BIO-W model

No.	Name	Category
1	Irrigated paddy rice	Crops
2	Rainfed paddy rice	
3	Irrigated wheat	
4	Rainfed wheat	
5	Irrigated coarse grain	
6	Rainfed coarse grain	
7	Irrigated oilseeds	
8	Rainfed oilseeds	
9	Irrigated sugar crop	
10	Rainfed sugar crop	
11	Irrigated other crops	
12	Rainfed other crops	
13	Forestry	Forestry
14	Dairy farms	Livestock & processed livestock products
15	Ruminant	
16	Non-ruminant	
17	Processed dairy	
18	Processed ruminant	
19	Processed non-ruminant	
20	Crude vegetable oil (producing crude vegetable oil and oilseeds meals)	Processed food products
21	Refined vegetable oil	
22	Beverage and sugar	
23	Processed rice	
24	Other processed food	
25	Processed feed	
26	Other primary sectors	Other primary sectors
27	Ethanol1 (producing ethanol and DDGS)	Bio-energy industries
28	Ethanol2	
29	Biodiesel	
30	Vegetable oil by-product	
31	Coal	Other energy industries
32	Oil	
33	Gas	
34	Oil Products	
35	Electricity	
36	Energy intensive industries	
37	Other industrial sectors	Rest of industries
38	Non-tradable service	

Table 3.5 Water scarcity shock (%)

Region	B1	B2	B3	B4	B5	B6	B7	B8	B9	B10
USA	–	–	–	–	–	10.1	–	–	–	–
EU27	–	–	0.7	–	–	18.1	–	–	–	–
BRAZIL	–	–	−16.1	–	–	–	−50	45.4	–	–
CAN	–	–	–	–	–	–	–	–	–	–
JAPAN	–	–	–	–	–	–	–	–	–	–
CHIHKG	−23	−13.1	−1	−2	−64.3	−24	−32	–	13.1	–
INDIA	−0.1	–	–	−10.6	−22	−9.2	2.9	−2.6	−7.9	−20
Central America	19.1	–	11.8	–	–	–	–	−2.3	–	–
South America	–	–	–	–	–	–	–	–	−7.9	–
East Asia	−10.1	–	–	–	–	–	–	–	–	–
MLYS & IDN	5.1	–	–	–	−0.4	–	–	–	–	–
R. Southeast Asia	–	13.5	−0.1	–	−26.7	−2.2	–	–	–	–
R. South Asia	−0.6	−2	−1	−43	–	–	–	–	–	–
Russia	–	−58.4	–	–	–	–	–	–	−14.2	−8
E-Europe-RFSU	−0.6	–	−29.7	12.3	–	−0.2	18.1	–	–	−11.7
R. Europe	–	–	–	–	–	–	–	–	–	–
M-East-N-Africa	−4.8	−34.6	−30.1	–	−4.3	–	–	−0.4	–	–
SSA	−1.2	–	−84.5	−20.5	–	−1.1	−2.4	−26.6	−3	–
Oceania	–	–	–	–	–	−24.4	15	–	–	–
Region	B11	B12	B13	B14	B15	B16	B17	B18	B19	B20
USA	–	–	–	–	–	–	–	–	–	0

(continued)

Table 3.5 (continued)

Region	B11	B12	B13	B14	B15	B16	B17	B18	B19	B20
EU27	–	−2	−11.4	–	–	–	–	–	–	0
BRAZIL	–	–	–	–	–	–	–	–	–	0
CAN	–	–	–	–	–	–	–	–	–	0
JAPAN	–	–	–	–	–	–	–	–	–	0
CHIHKG	–	–	−0.6	−15.5	−0.1	–	–	−23	−13.1	0
INDIA	–	−61.3	–	–	–	–	–	−0.1	–	0
Central America	–	–	–	–	–	–	–	19.1	–	0
South America	−15	–	–	–	–	–	–	–	–	0
East Asia	–	–	–	–	–	–	–	−10.1	–	0
MLYS & IDN	–	–	–	–	–	–	–	5.1	–	0
R. Southeast Asia	–	–	–	–	–	–	–	–	13.5	0
R. South Asia	–	–	–	–	–	–	–	−0.6	−2	0
Russia	−0.1	−16.2	–	–	–	–	–	–	−58.4	0
E-Europe-RFSU	–	−0.3	−13.8	−8.3	−30.3	–	–	−0.6	–	0
R. Europe	–	–	–	–	–	–	–	–	–	0
M-East-N-Africa	–	–	–	–	–	–	–	−4.8	−34.6	0
SSA	–	–	−0.7	−21	−0.2	9.8	–	−1.2	–	0
Oceania	–	–	–	–	–	–	–	–	–	0

Note Values are percentage point change of IWSR from 2000 to 2030. Negative numbers indicate potential water demand will be less satisfied by actual water consumption in 2030 than in 2000

Table 3.6 A comparison of the welfare change across models, all regions included

	Full model	Combined I&R model	Unified basin model	Single-region-Brazil	Single-region-China	Single-region-India	Single-region-RSouthAsia	Single-region-MENA	Single-Region-SSA
NAmr	542	1139	487						
Eurp	−725	−58	−603						
Brazil	128	75	112	2					
LAmr	199	128	174						
China	−995	−2358	−1006		−797				
India	−784	−7184	−789			−751			
RSouthAsia	−1191	−4124	−1119				−1133		
RAsia	−706	−2368	−604						
MENA	−481	−719	−395					−142	
SSA	68	−622	51						−19
OCN	220	318	196						
World	−3725	−15,773	−3496						

Unit is million USD 2011

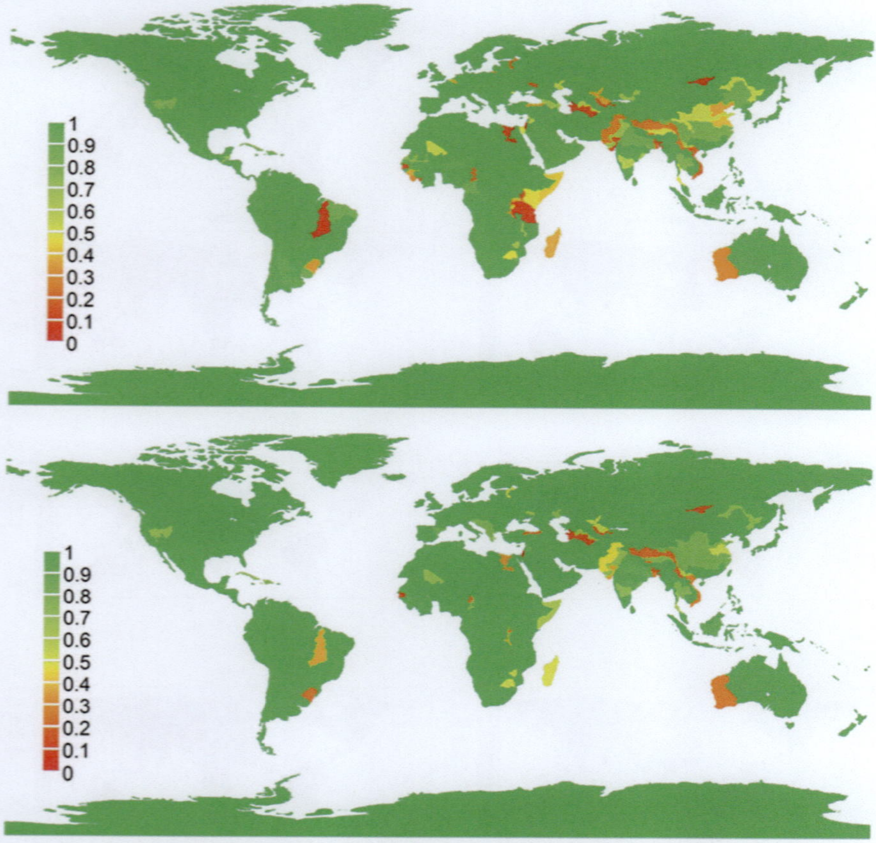

Fig. 3.7 Evolving irrigation water supply reliability (Year 2030 (top) relative to year 2000 (bottom)). The index value of one indicates that all the potential irrigation demand is satisfied by the actual irrigation water consumption. A smaller index value indicates more severe irrigation water shortfall.)

as the fundamental unit of water balance. It simulates reservoir regulation of natural flow and abstraction of surface and groundwater based on projected total water demand for domestic, industrial, livestock and irrigation sectors. Irrigation water demand is estimated using effective rainfall and potential evapotranspiration generated by the IGHM, plus irrigated areas, cropping patterns, crop characteristics, and basin irrigation efficiency. With projected sectoral water demand, the IWSM optimizes water supply according to demand, subject to water availability and capacity constraints of water infrastructure. Sequentially, the model first calculates total monthly water supply; second, it allocates the total supply to water-use sectors on a priority-based manner, assuming domestic water demand is the first priority, industrial and livestock demand is the second priority, and the remaining water is available

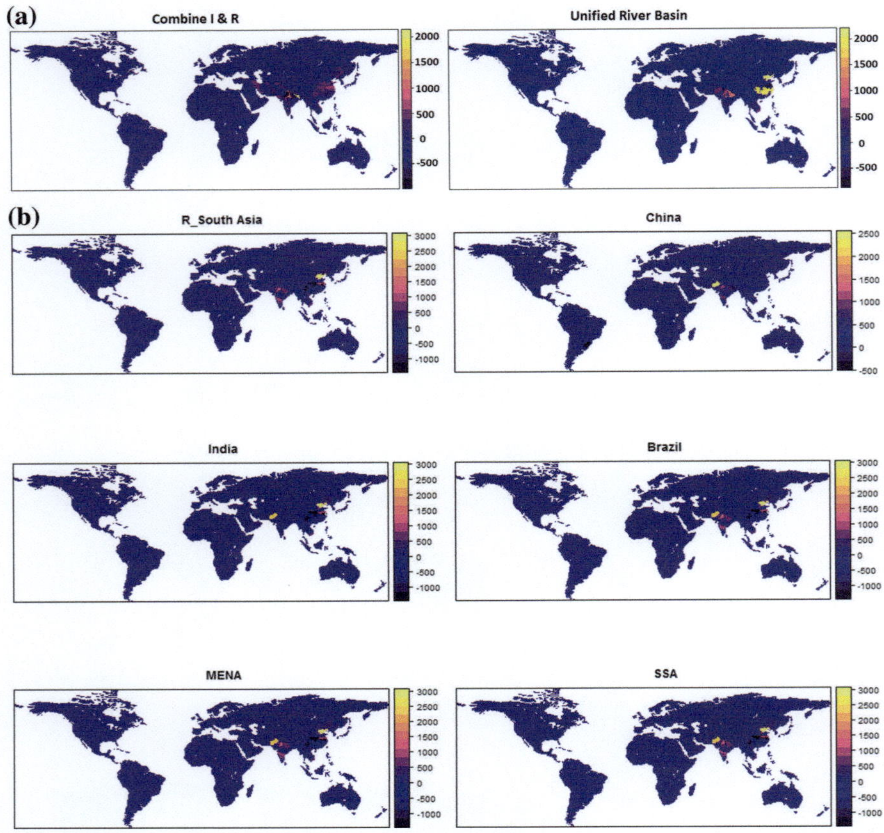

Fig. 3.8 Deviations of irrigated land use change (in 1000 ha) from the benchmark model. (Top panel (**a**) shows the deviations produced by the combined I&R model and the unified basin model. Bottom panel (**b**) shows the deviations produced by the respective single-region models.)

for irrigation. Total irrigation water supply is further allocated to crops according to crop water requirements. As a water scarcity indicator, irrigation water supply reliability is determined as the ratio of total irrigation water supply to demand, on an annual basis.

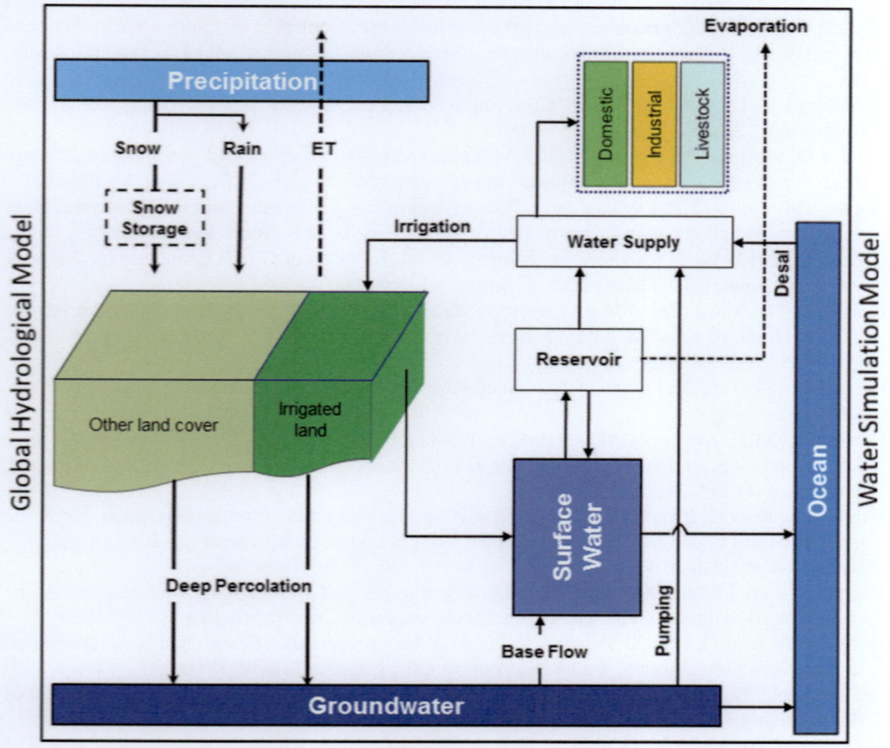

Fig. 3.9 Structure of the global hydrological model IGHM and water simulation model IWSM. Adapted from Zhu et al. (2013)

References

Berrittella M, Hoekstra AY, Rehdanz K, Roson R, Tol RSJ (2007) The economic impact of restricted water supply: a computable general equilibrium analysis. Water Res 41:1799–1813

Cai X, Rosegrant MW (2002) Global water demand and supply projections part 1: a modeling approach. Water Int 27:159–169

Cakmak E, Dudu H, Saracoglu S (2009) Climate change and agriculture in Turkey: a CGE modeling approach. In: EconAnadolu 2009: Anadolu international conference in economics

Calzadilla A, Rehdanz K, Tol RSJ (2010) The economic impact of more sustainable water use in agriculture: a computable general equilibrium analysis. J Hydrol 384:292–305

Decaluwe B, Patry A, Savard L (1999) When water is no longer heaven sent: comparative pricing analysis in a AGE model. Département d'économique, Université Laval Working Paper 9908

Diao X, Roe T (2003) Can a water market avert the "double-whammy" of trade reform and lead to a "win–win" outcome? J Environ Econ Manag 45:708–723

Dinar A (2014) Water and economy-wide policy interventions. Found Trends R Microecon 10:85–165

Dixon PB, Rimmer MT, Wittwer G (2011) Saving the Southern Murray-darling basin: the economic effects of a buyback of irrigation water. Econ Rec 87:153–168

Dudu H, Chumi S (2008) Economics of irrigation water management: a literature survey with focus on partial and general equilibrium models. SSRN Scholarly Paper No. ID 1106504. Rochester, NY

Falkenmark M, Lannerstad M (2005) Consumptive water use to feed humanity - curing a blind spot. Hydrol Earth Syst Sci 9:15–28

Garrick D, Whitten SM, Coggan A (2013) Understanding the evolution and performance of water markets and allocation policy: a transaction costs analysis framework. Ecol Econ 88:195–205

Gomez CM, Tirado D, Rey-Maquieira J (2004) Water exchanges versus water works: insights from a computable general equilibrium model for the Balearic Islands. Water Resour Res 40

Griffith M (2012) Water resources modeling: a review. In: Wittwer G (ed) Economic modeling of water: the Australian CGE experience. Springer, Dordrecht, pp 59–77

Hassan R, Thurlow J (2011) Macro–micro feedback links of water management in South Africa: CGE analyses of selected policy regimes. Agric Econ 42:235–247. https://doi.org/10.1111/j.1574-0862.2010.00511.x

Hertel TW (1997) Global trade analysis: modeling and applications. Cambridge University Press, New York

Hertel TW, Golub AA, Jones AD, O'Hare M, Plevin RJ, Kammen DM (2010) Effects of US Maize ethanol on global land use and greenhouse gas emissions: estimating market-mediated responses. Bioscience 60:223–231

Kahsay TN, Kuik O, Brouwer R, van der Zaag P (2015) Estimation of the transboundary economic impacts of the Grand Ethiopia Renaissance Dam: a computable general equilibrium analysis. Water Resour Econ 10:14–30

Keeney R, Hertel TW (2009) The indirect land use impacts of United States biofuel policies: the importance of acreage, yield, and bilateral trade responses. Am J Agric Econ 91:895–909

Koopman JF, Kuik O, Tol RSJ, Brouwer R (2015) Water scarcity from climate change and adaptation response in an international river basin context. Clim Change Econ 06:1550004

Liu J, Hertel TW, Taheripour F, Zhu T, Ringler C (2014) International trade buffers the impact of future irrigation shortfalls. Glob Environ Change 29:22–31

Luckmann J, Grethe H, McDonald S, Orlov A, Siddig K (2014) An integrated economic model of multiple types and uses of water. Water Resour Res 50:3875–3892

Maupin MA, Kenny JF, Hutson SS, Lovelace JK, Barber NL, Linsey KS (2014) Estimated use of water in the United States in 2010. USGS No. 1405. U.S. Geological Survey, Reston, VA

Monfreda C, Ramankutty N, Foley JA (2008) Farming the planet: 2. Geographic distribution of crop areas, yields, physiological types, and net primary production in the year 2000. Glob Biogeochem Cycles 22

Olmstead SM (2014) Climate change adaptation and water resource management: a review of the literature. Energy Econ 46:500–509

Ponce R, Bosello F, Giupponi C (2012) Integrating water resources into computable general equilibrium models - a survey. Fondazione Eni Enrico Mattei Work, Pap

Portmann FT, Siebert S, Döll P (2010) MIRCA2000—Global monthly irrigated and rainfed crop areas around the year 2000: a new high-resolution data set for agricultural and hydrological modeling. Glob Biogeochem Cycles 24

Ramankutty N, Evan AT, Monfreda C Foley JA (2008) Farming the planet: 1. Geographic distribution of global agricultural lands in the year 2000. Glob Biogeochem Cycles 22

Rosegrant MW, Ringler C, Zhu T, Tokgoz S, Bhandary P (2013) Water and food in the bioeconomy—challenges and opportunities for development. Agric Econ 44:139–150

Siebert S, Döll P (2010) Quantifying blue and green virtual water contents in global crop production as well as potential production losses without irrigation. J Hydrol 384:198–217

Taheripour F, Hertel T, Liu J (2013) Introducing water by river basin into the GTAP-BIO model: GTAP-BIO-W. GTAP Work. Pap. No 77. http://www.gtap.agecon.purdue.edu/resources/res_display.asp?RecordID=4304. Accessed 27 Sept 2018

Wittwer G (ed) (2012) Economic modeling of water: the Australian CGE experience. Springer, Dordrecht, New York

Young RA (1986) Why are there so few transactions among water users? Am J Agric Econ 68(5):1143–1151

Zhu T, Ringler C, Iqbal MM, Sulser TB, Goheer MA (2013) Climate change impacts and adaptation options for water and food in Pakistan: scenario analysis using an integrated global water and food projections model. Water Int 38(5), pp.651-669.

Dr. Jing Liu is Research Economist at the Department of Agricultural Economics, Purdue University. She works on issues at the interface between humankind and environment. Her research seeks to improve our understanding of the economic impacts of environmental changes and our ability to adapt to those impacts. Dr. Liu specializes in using computational and statistical tools to analyze interactions between food production system, land and water. She has also worked across a spectrum of applied economic topics including economic damage of natural disasters, international trade, as well as health and nutrition.

Dr. Tom Hertel is Distinguished Professor of Agricultural Economics at Purdue University, where his research and teaching focus on international trade, food and environmental security. Dr. Hertel is a Fellow, and a Past-President, of the Agricultural and Applied Economics Association (AAEA). He is also the founder and Executive Director of the Global Trade Analysis Project (GTAP). The original project model and database is documented in the book: Global Trade Analysis: Modeling and Applications, edited by Dr. Hertel, and published by Cambridge University Press. He has supervised more than forty Ph.D. students and published more than 120 peer reviewed journal articles, along with several dozen book chapters as well as four books. Professor Hertel is the inaugural recipient of the Purdue University Research and Scholarship Distinction Award. He has also received a number of AAEA awards including: Publication of Enduring Quality, Distinguished Policy Contribution, Outstanding Journal Article and Quality of Communication. He has also been Advisor to two Outstanding AAEA Ph.D. and MS theses.

Dr. Farzad Taheripour is an associate professor in energy economics in the Department of Agricultural Economics at Purdue University. He received his Ph.D. in Agricultural Economics from the University of Illinois at Urbana-Champaign in 2006. Dr. Taheripour's research interests are in energy, agriculture, policy analysis and economic modeling. He has over 120 professional publications including journal papers, book chapters, conference papers, and reports. He collaborates with several national and international organizations and institutions. He is a leading researcher in the area of economic modeling and made major contributions in developing multiple versions of the GTAP model. He is a GTAP Research Fellow for the term of 2017–2020.

Chapter 4
Beyond Water Stress: Structural Adjustment and Macroeconomic Consequences of the Emerging Water Scarcity

Roberto Roson

Abstract This work analyzes some system-wide macroeconomic consequences of lower (sustainable) water availability, when global economic growth is postulated according to the Shared Socio-Economic Pathway 1 (SSP1), for the reference year 2050. After finding that the rather optimistic forecasts of economic development cannot be met in most water scarce macro-regions, we assess what consequences for the structure of the economy, welfare and the terms of trade, the insufficiency of water resources would imply. The analysis is undertaken by means of numerical simulations with a global computable general equilibrium model, under a set of alternative hypotheses. In particular, we consider whether (or not) the regional economic systems have a differentiated capability of adaptation (by means of innovation and modification of economic processes), and whether (or not) the scarce water resources can be allocated among industries, such that more water is assigned where its economic value is greater.

Keywords Water · Economic growth · Shared socio-economic pathways · Computable general equilibrium · Virtual water trade

4.1 Introduction

Will economic development be hampered by a lack, or by an unsustainable exploitation, of water resources? For almost all studies addressing the issue of water availability from a macroeconomic perspective (e.g., Rosegrant et al. 2002), the answer is yes. The world will face more frequent and more severe water crises in the future, with

R. Roson (✉)
Ca' Foscari University, Venice, Italy
e-mail: roson@unive.it

R. Roson
IEFE Bocconi University, Milan, Italy

R. Roson
Loyola Andalusia University, Seville, Spain

G. Wittwer (ed.), *Economy-Wide Modeling of Water at Regional and Global Scales*, Advances in Applied General Equilibrium Modeling,
https://doi.org/10.1007/978-981-13-6101-2_4

negative and serious socio-economic implications, especially for some developing countries, already hit by climate change impacts.

A largely ignored question is, however, how the emerging water scarcity will shape the economy in the medium and long term. In other words: facing an ever increasing (explicit or implicit) cost for water, how the production and consumption processes will change? What the ultimate welfare effects will be? How much economic growth potential will be lost?

Clearly, possible answers to all the issues above are not deterministic. Much depends on public policies, but also on individual choices, as well as on the technological options available. Broadacre farmers have been using minimum tillage practices for more than three decades, resulting in improved moisture retention and soil conservation. Biotechnology plays an ongoing role in improving the drought tolerance of crops (Deikman et al. 2012). Other technologies are reducing water requirements (for example, Wilson 2018). Here we explore some scenarios through numerical simulations with a global computable general equilibrium model, under a set of alternative hypotheses. In particular, we consider whether (or not) the regional economic systems have a differentiated capability of adaptation, and whether (or not) the scarce water resources can be allocated among industries, such that more water is assigned where its economic value is greater.

This research adds to the literature by directly addressing the feedback from water scarcity to the economic system. The conventional approach goes to the opposite direction: a potential water demand is first assessed and then compared with some measure of supply/availability, often by constructing an index of "water stress" (e.g., Alcamo et al. 2003; Arnell 2004). This does not inform us about how the latent excess of demand for water could be absorbed.

In the following section, the modelling strategy and the design of simulation exercises is illustrated. Results are presented in Sect. 4.3, followed by a discussion about the significance and limitations of our findings. After the conclusions, three Appendices provide additional information on some technical aspects of our methodology.

4.2 The Design of a Set of Simulation Exercises

We base our modeling exercise on the Shared Socio-Economic Scenario 1 (SSP1), referred to the year 2050 (Kriegler et al. 2012; O'Neill et al. 2014), for projections of GDP and population. SSP1 is characterized by the following narrative: "Sustainable development proceeds at a reasonably high pace, inequalities are lessened, technological change is rapid and directed toward environmentally friendly processes, including lower carbon energy sources and high productivity of land".

As in Roson (2016) and Roson and Damania (2017) CGE simulations are conducted to extrapolate the potential water demand, consistent with the economic growth hypotheses of SSP1, as well as to assess the amount of future water deficits (excess demand for water), occurring at a given year. The whole process is described in more detail in the Appendix 1.

Table 4.1 Percentage[a] of potential water demand exceeding sustainable supply

	SSP1-2050
MENA	−77.1
C_Asia	−4.4
E_Asia	−42.9
S_Asia	−50.2

[a]Percentages in this table are (PD/SS − 1)*100, where PD is "potential demand" and SS "sustainable supply", broadly corresponding to yearly total runoff in the region

Our global model considers 14 macro-regions, and we found that potential water demand would exceed "sustainable" levels of aggregate water withdrawal in four of them: Middle East and North Africa (MENA), Central (C_Asia), East (E_Asia) and South Asia (S_Asia), at varying degrees, as shown in Table 4.1.

If water availability will become a constraint for the regional economies, scenarios of economic growth (which is especially sustained in the SSP1), and their associated potential water demand, will turn out to be incompatible with the actual availability of resources.

How the excess demand for water would be absorbed by the economic system? First, efficiency improvements could be achieved by reallocating water consuming activities in time and space, within a region and during the year. This can be obtained through the normal market functioning, driven by the changing relative competitiveness of the various economic units, induced by water scarcity. For instance, sowing times and crops can be changed in agriculture, new energy plants can be built where water supply is sufficient and stable, etc. Where water rights markets are developed, water allocations can be banked and traded. Second, water saving technologies, policies and processes could be introduced, as the corresponding investment would be justified by the rising cost of water. For example, pipelines can be built for water transfer projects, as well as infrastructure for desalination and wastewater recycling.

It is extremely difficult to gauge the degree of endogenous efficiency gains potentially achievable in the water stressed macro-regions, at the spatial and temporal scale adopted in our global model. To conduct our simulations experiments, we therefore adopt two sets of assumptions. The first one is a simple benchmark case, consistent with our previous work (Roson and Damania 2017), were it is assumed that efficiency improvements can cover 75% of the demand gap in all water stressed regions. The remaining 25% is interpreted as cuts in water availability; for instance, the 77.1% excess demand in the MENA region would bring about a reduction in water availability of 19.27%. Alternatively, we consider various factors (economic, technical and institutional) that could ultimately affect the actual degree of "flexibility" or "absorption capacity" in the regional economic systems. Appendix 3 illustrates how a scenario of regionally differentiated impacts, based on a qualitative index, has been built. In this alternative setting, we assume that the reductions in water availability, expressed as a share of the regional demand gap (25% in the base case), are fixed at 26% for MENA, 34% for C_Asia, 8% for E_Asia, 18% for S_Asia.

The reductions in water availability, determined for the regional aggregate, determine reductions at the finer sectoral level. Again, we use here two settings. The first one is a simple benchmark where sectoral water availability is reduced proportionally across the board. For example, −19.27% water in the MENA region corresponds to −19.27% in all water-using sectors inside the MENA. The second case is a little more elaborated, because sectoral reductions are made sensitive to the relative water efficiency. As in Roson and Damania (2017) sectoral reductions (ρ_{ir}) are increasing functions of the "water per unit of output", or water intensity, coefficients (ω_{ir}):

$$\rho_{ir} = \alpha_r + \left(\frac{\omega_{ir}}{\overline{\omega}_r}\right)^{0.25} \tag{1}$$

where α_r is a constant parameter, determined for each region r, set at a level ensuring that the sum of all sectoral reductions matches the one imposed to the regional aggregate, and $\overline{\omega}_r$ is the average water intensity in the region. Roson and Damania (2017) explain how these water intensity coefficients have been estimated. The inverse of the water intensity can be interpreted as the (average) water productivity. Therefore, (1) establishes that sectoral reductions in water availability are larger for those sectors with lower relative water efficiency, typically in agriculture, and vice versa.

Combining the alternative assumptions, we obtain three scenarios for our numerical simulations:

- No regional differentiation in absorption capacity, uniform reduction of water availability in all sectors [NRUS];
- Regional differentiation in absorption capacity (see Appendix 3), uniform reduction of water availability in all sectors [DRUS];
- Regional differentiation in absorption capacity (see Appendix 3), non-uniform (efficiency sensitive) reduction of water availability in the various sectors (see Appendix 2) [DRES].

By selecting one of the three cases above, one implicitly determines how large the cuts in water consumption are for each regional industry. These are subsequently translated in terms of changes in (multifactor) industrial productivity. For example, less water in agriculture implies lower yields, *ceteris paribus*. In this study, the water-induced variation in productivity depends on specific characteristics of the different industries, which are captured by a set of "water-output elasticity" parameters. Appendix 2 illustrates how region and industry specific parameters for the output elasticity and the marginal value of water have been estimated. The mean output elasticity is 0.8. This implies that, on average, a 10% reduction in water usage entails an 8% reduction in the gross industrial output, if other production factors are not adjusted.

Linking water to productivity allows us to conduct some numerical simulations with the global GTAP CGE model (Hertel and Tsigas 1997), because productivity parameters are exogenous in that system, in the equilibrium conditions for the various markets. After changing some productivity factors, then, a new counterfactual equilibrium is computed, where variations in endogenous macroeconomic variables, like relative prices, trade flows, production volumes and others can be analyzed. Clearly, the purpose of the simulations is not that of producing forecasts, but to isolate the systemic causal effects of a specific shock (in our case, water-induced productivity variations), from the many factors which could ultimately affect the economic system.

In the following section, some findings corresponding to the three simulation cases are presented and discussed. The results are expressed as relative to a hypothetical reference, where the regional economies grow at the rate imposed by the SSP1, but there is no lack of water resources, thus no effects on productivity related to water availability. Therefore, a negative variation of—for instance—regional income should not be literally interpreted as a reduction, but rather as a growth rate smaller than the one hypothetically set by the SSP1.

4.3 Results

Variations in water availability for each industry in water stressed macro-regions are defined for the three simulation scenarios (NRUS, DRUS, DRES) and transformed as exogenous productivity shocks in the CGE model. For example, Table 4.2 shows the productivity variations for MENA in the three cases.

We start by considering the variations in Real GDP, or national income, relative to the benchmark where water scarcity has no impact on industrial productivity (Table 4.3).

Notice that, when water efficiency is considered in the allocation scheme DRES, a few industries obtain productivity gains. This is because, even if water consumption is reduced in the region as a whole, some individual industries, where relatively little water is used per unit (value) of output, actually get increases in water assignments.

The CGE system computes a global economic equilibrium consistent with the exogenous shocks above, and the model delivers estimates for several macroeconomic variables, like: production volumes, employment, investments, consumption patterns, trade flows, price indexes, GDP deflators, etc. We present here only a limited set of results, to illustrate the key characteristics of the three scenarios.

Drops in (potential) GDP are quite substantial in water stressed macro-regions, most notably in the Middle East and North Africa (MENA). When differences in absorption capacities are taken into account, the picture changes significantly for East Asia (dominated by China), because we are assuming that as much as 92% of the water deficit can be accommodated there, through endogenous efficiency gains and technological progress.

Results for the DRES scenario indicate that the economic impact of water scarcity can be greatly contained when the economic returns per unit of water are considered in the industrial rationing scheme. Interestingly, the negative shock turns positive in the Asian regions, where the aggregate efficiency gains of a better inter-industrial allocation of water resources overrules the direct productivity effect (Table 4.3).

Table 4.2 Industrial productivity shocks in the MENA region

	NRUS (%)	DRUS (%)	DRES (%)
Rice	−35.74	−37.17	−70.42
Wheat	−24.03	−25.00	−45.26
Cereals	−24.27	−25.25	−47.36
VegFruit	−30.73	−31.96	−40.63
Oilseeds	−21.67	−22.54	−37.89
Sugar	−22.34	−23.24	−34.63
Oth Crops	−18.06	−18.78	−17.20
Oth Agr.	−36.20	−37.65	−35.17
Extr	−18.82	−19.57	−13.10
P.Food	−14.03	−14.60	5.30
Textiles	−17.29	−17.98	7.05
Light Man	−23.21	−24.13	9.63
Heavy Man	−13.79	−14.35	3.75
Utilities	−13.38	−13.91	−7.59

(These are the reductions in gross output when water is cut and other production factors are not adjusted. For instance, the −70.4% for Rice in DRES comes because (a) rice production is sensitive to water availability, (b) water is substantially reduced in the MENA, and (c) water is diverted disproportionally away from rice production with rising scarcity)

Table 4.3 Variations in real GDP

	NRUS (%)	DRUS (%)	DRES (%)
N_America	−0.01	−0.01	0.00
C_America	0.05	0.05	0.06
S_America	−0.01	0.00	0.01
W_Europe	0.00	0.00	−0.01
E_Europe	−0.04	−0.03	−0.05
MENA	−8.64	−8.99	−3.77
Sahel	0.21	0.28	0.51
C_Africa	0.26	0.27	0.29
S_Africa	−0.02	0.00	0.05
C_Asia	−0.38	−0.57	0.42
E_Asia	−2.47	−0.80	0.42
S_Asia	−4.17	−3.36	0.44
SE_Asia	−0.03	0.00	0.01
Australasia	−0.02	−0.01	0.01

Table 4.4 EV/GDP

	NRUS (%)	DRUS (%)	DRES (%)
N_America	−0.03	−0.02	0.00
C_America	0.01	0.02	0.02
S_America	0.01	0.01	0.02
W_Europe	−0.03	−0.02	−0.01
E_Europe	0.01	0.01	0.01
MENA	−1.18	−1.23	−0.43
Sahel	0.07	0.08	0.10
C_Africa	0.16	0.15	0.14
S_Africa	0.02	0.02	0.03
C_Asia	0.09	0.07	0.14
E_Asia	−0.14	−0.05	0.01
S_Asia	−0.25	−0.21	0.04
SE_Asia	0.02	0.02	0.02
Australasia	0.00	0.00	0.00

To appreciate the impacts on the welfare of households, we employ the Equivalent Variation (EV) concept. The EV is the welfare equivalent reduction in income, virtually obtainable at unchanged relative prices. In other words, it measures the reduction in income, which would lower the average household welfare the same way as water scarcity would do. This can be interpreted as the welfare "cost" of water scarcity in the various scenarios. Table 4.4 expresses the EV as relative to the benchmark GDP level.

Table 4.4 closely mirrors Table 4.3, but with much smaller figures in absolute value. The reason is that water scarcity affects regions and industries in a differentiated way, so that the representative regional consumer can substitute water-intensive and relatively more expensive goods with cheaper ones, possibly imported from abroad. This endogenous substitution mechanism curbs the overall impact on the welfare of consumers.

Correspondingly, the industrial structure changes in the regions where water gets progressively scarcer. By way of illustration, Table 4.5 presents the changes in industrial output volumes in the MENA region.

In the Middle East and North Africa, we can see that the industries most vulnerable to reductions in water availability are found in Agriculture, with the addition of Extraction and Manufacturing. However, when water is reallocated on the basis of economic returns, agricultural productions like rice, wheat, oil seeds and others are significantly cut down, whereas the production of textiles and manufactured goods is expanded.

Using terminology from international trade theory, we can say that MENA gets a comparative disadvantage in water-intensive industries, when water gets scarce and affects productivity. In other words, it is not economically efficient to produce water

Table 4.5 Industrial output changes in the MENA region

	NRUS (%)	DRUS (%)	DRES (%)
Rice	−12.83	−13.69	−26.70
Wheat	−14.31	−15.20	−31.37
Cereals	−9.25	−9.70	−15.03
VegFruit	−6.31	−6.67	−8.20
Oilseeds	−15.67	−16.61	−33.38
Sugar	−6.02	−6.30	−4.45
Fibers	−7.95	−8.50	−5.61
Oth Crops	−13.32	−14.24	−26.39
Oth Agr.	−12.93	−13.80	−13.91
Extraction	−18.32	−19.26	−14.40
P.Food	−6.96	−7.26	−4.09
Textiles	−7.49	−8.39	7.73
Light Man	−18.80	−19.90	13.08
Heavy Man	−12.22	−12.86	3.66
Electricity	−7.96	−8.30	−0.92
GasDis	−8.90	−9.32	−3.56
WaterDis	−5.95	−6.11	−2.89
Construction	−2.76	−2.65	−0.48
TransComm	−2.43	−2.29	−0.51
OthServices	−1.83	−1.76	−1.15

intensive crops like rice or oil seeds, as long as they can be safely imported and paid with exports (e.g., light manufacturing), having a lower impact on water resources.[1]

Another way to highlight the same effect is through the virtual water concept. Since water is needed for the production of goods and services, the water employed is virtually exported as "embodied" into the traded goods. Analogously, importing goods (and services) requiring water in their production processes can be interpreted as virtual water imports.

Since we have data on water utilized in all industries and regions, we can readily estimate regional "virtual water trade balances". A virtual water trade deficit can be interpreted as saved (domestic) water resources, because of trade exchanges with the rest of the world. Furthermore, we can assess whether and how much the virtual water balance varies in the three simulation scenarios, when water consumption is reduced in some regions. The variations in the virtual water trade balances are displayed in Table 4.6.

Our results suggest that international trade is an important mechanism to alleviate the economic consequences of water scarcity, especially for the MENA. However,

---

[1]Another example is Australia, which, during severe droughts, may resort to importing feed grain when it normally does not.

Table 4.6 Variations in virtual water trade balances (millions m^3)

	NRUS	DRUS	DRES
N_America	5657	4365	6951
C_America	1329	1213	1479
S_America	2292	1911	3877
W_Europe	3306	2996	3687
E_Europe	4491	4295	8116
MENA	−23,393	−24,764	−40,979
Sahel	150	54	292
C_Africa	−93	−212	192
S_Africa	1086	1002	1547
C_Asia	407	265	980
E_Asia	2907	6117	6046
S_Asia	−3116	−1410	795
SE_Asia	3924	3420	5024
Australasia	1052	747	1993

there is no simple and direct relationship between virtual water trade and water availability. This is because virtual water trade reflects real trade in tangible goods, especially agricultural products, and the structure of international trade is actually affected by several factors, not just the degree of access to natural resources (Reimer 2012).

4.4 Discussion

To correctly interpret the findings above, one should consider the limitations of our analysis, imposed by its geographical and temporal scale. Whereas detailed studies are available, which take into account (usually for a given river basin) several physical and human processes operating at a daily, weekly or monthly period, our global macroeconomic model is based on official national accounts, registering market transactions in a year. Since water is not formally traded in many economies, or its price is significantly distorted, it is not generally possible to ascertain the contribution of water as a production factor in the various industries. Furthermore, we are considering very large macro-regions, and the national boundaries do not overlap with river basins, nor include territories homogeneous in terms of hydrologic, climatic and physical conditions in general.

This means that we are unable to capture and realistically model the many adjustment processes occurring inside an economic system facing rising degrees of water scarcity. On the other hand, our analysis provides a much broader picture of the systemic effects of water scarcity, at a global scale, which simply cannot be offered by more detailed but narrowly focusing studies.

In order to introduce the various adjustment processes and to model the impacts of water scarcity in the economy, we therefore rely on a limited set of data, supplemented by informed guesses and specific assumptions, which are of course subject to debate. This choice was not only imposed by necessity, as we understand that adaptation behavior and policies do not necessarily follow a strictly rational economic logic.[2]

All in all, we come to the rather paradoxical outcome that our analysis, despite being based on a mathematical model providing many detailed numerical results, should be viewed as a qualitative assessment, rather than as a purely quantitative analysis. Indeed, the ultimate purpose of this study is providing an order of magnitude for the effects at play, while highlighting the complex structural adjustment mechanisms, triggered by the emerging water scarcity in some regions of the world.

4.5 Conclusion

Global modelling scenarios are increasingly taking into account the constraints associated with natural resources availability. This study focuses on water, pointing out that there will be an emerging issue of resource availability in some water scarce (and mostly developing) countries, which, because of this, may not be able to fully reap their economic development potential.

Adaptation to the emerging water scarcity will involve several adjustments in the consumption and production patterns, trade flows, income and welfare levels, not only in the regions and sectors initially affected by the shortage of water resources, but in all interconnected global markets and economies.

The eventual economic impact of water scarcity will depend on several factors, and a number of policy options are available to curb its negative consequences. However, no option will come without a cost (in a broad sense, including political acceptability and effort). The overall flexibility of the economic systems will be very important, as well as the introduction of technologies and processes aimed at improving water efficiency levels.

Our study also reveals that a better allocation of scarce water resources among sectors could play a pivotal role in this respect, suggesting that the issue of water scarcity is not merely a physical but also an economic problem. Accurate estimates of the economic value of water will therefore be essential for a rational allocation of scarce water across locations, uses, users, and time periods (Ward and Michelsen 2002).

Acknowledgements Richard Damania, Glyn Wittwer and Ruslana Palatnik provided useful comments and suggestions on some earlier versions of this work. The usual disclaimer applies.

---

[2] A good example is provided by the analysis of water scarcity mitigation options in Israel provided by Becker et al. (2010), making clear that innovation in water systems also entails complicated political processes. In addition, learning and knowledge diffusion matter. For instance, rice farmers in Australia were most reluctant to trade water when water reforms first made it possible. But in the drought of 2002–03, they found it advantageous to sell water for around $300/ML when the average product of water in their production was probably under $250/ML.

Appendix 1: An Assessment of Future Water Deficits

Although this study focuses only, as an illustrative case, on the year 2050 and the SSP1 scenario, future potential water deficits were estimated for other years and another scenarios (see Roson and Damania 2017).

The assessment of the future water deficits has been based on a limited set of SSP forecasts of income and population growth, complemented by CGE simulations aimed at enlarging the number of estimated economic variables. For each combination of year and SSP, growth rates in population and GDP have been assumed, using data from the IIASA SSP repository. By shocking the corresponding parameters in the GTAP CGE model (dataset 9.0), several other endogenous variables were obtained, like production volumes by industry and region, household consumption, regional investments, exports and imports, income by source, etc.

Estimates of industrial output are especially relevant because, coupled with some econometrically computed future water intensity coefficients, allow to derive the implied water demand. Analogously, municipal water demand was computed by assuming it dependent on population growth, real income levels and a trend of increased water efficiency. Table 4.7 presents the water demand projections for the SSP1 at the year 2050.

Regional water deficits are defined as the difference between potential water demand and sustainable water supply. In turn, the latter is identified as the sum of water runoff and inflow in a region, estimated by the global hydrologic GCAM model,[3] driven by three different Global Circulation Models (CCSM, GISS, FIO ESM). We found that four macro-regions have levels of potential SSP1 demand exceeding sustainable supply in the year 2050.

Appendix 2: Estimation of the Marginal Value and Output Elasticity of Water

When water is regarded as a production factor, the Marginal Value of Water (MVW) is the increase in the value of output potentially obtainable when one unit of water (here, one cubic meter) is added to the process, while keeping all the other production factors unchanged. The concept is strictly linked to that of water pricing and allocation: (a) profit maximization and cost minimization imply that MVW should equate the price of water; (b) water (or any other resource) is efficiently allocated (from an economic viewpoint) when its marginal value is the same across alternative uses.

In principle, estimating the MVW would require specific technical information on the production processes and how water contributes to them. This is simply impossible to get for large aggregate sectors and regions. Instead, we propose here a methodology for a consistent estimation of MVW in 15 industries and 14 macro-regions,

[3]http://www.globalchange.umd.edu/models/gcam.

Table 4.7 Water demand projections (potential demand consistent with SSP scenario 1)

	1. N_America	2. C_America	3. S_America	4. W_Europe	5. E_Europe	6. MENA	7. Sahel	8. C_Africa	9. S_Africa	10. C_Asia	11. E_Asia
Baseline 2004											
Agriculture	1343	437	990	363	787	558	416	561	279	150	1722
Industrial	526	123	176	172	403	518	7	52	58	50	301
Municipal	39	26	18	16	29	29	3	3	6	5	80
Total	**1908**	**586**	**1184**	**551**	**1219**	**1106**	**426**	**616**	**343**	**205**	**2103**
2050 SSP1											
Agriculture	1559	588	1319	413	1053	773	779	1008	447	236	2987
(%)	16.1	34.5	33.2	13.9	33.7	38.5	87.1	79.6	60.5	57.3	73.5
Industrial	792	193	357	266	663	859	16	111	123	104	683
(%)	50.7	57.1	102.6	55.1	64.6	65.6	120.6	114.3	110.7	108.3	126.9
Municipal	66	59	44	26	58	83	22	25	32	23	396
(%)	69.8	130.7	145.2	58.0	100.6	182.3	680.8	661.6	426.5	346.1	393.8
Total	**2418**	**840**	**1719**	**705**	**1774**	**1714**	**817**	**1144**	**602**	**363**	**4066**
(%)	26.7	43.4	45.2	28.0	45.5	55.0	91.5	85.6	75.5	77.1	93.3
Δ GDP2050 (%)	**142.9**	**400.0**	**456.4**	**157.6**	**379.5**	**484.7**	**2160.8**	**2085.8**	**1341.6**	**1204.7**	**1426.4**

based on some available "water intensity coefficients" (WIC—water per value of output) and two calibrated parameters. WIC (indicated in the following as ω) and MVW are related but distinct concepts. Mathematically, WIC is just the ratio of water over output (in value terms), whereas MVW is the partial derivative of output value with respect to water.

The estimation procedure is based on a set of sensible assumptions one could impose on the water elasticity of output (ε). The latter is defined as the relative (percentage) variation of output (x) obtainable through a relative variation in the water input (w), ceteris paribus:

$$\varepsilon = \frac{\delta x/x}{\delta w/w} = \frac{\delta x}{\delta w}\frac{w}{x} = \frac{\delta x}{\delta w}\omega = MVW\omega \quad (2)$$

Consider ε to be a function of ω. Obviously, one would require that $\varepsilon(0) = 0$, because no variation in output would be observed if water is not used at all. A second sensible assumption is:

$$\lim_{\omega \to \infty} \varepsilon(\omega) = 1 \quad (3)$$

Meaning that, as water becomes the only relevant factor (enormous amounts of water are employed), the output varies proportionally with water (constant returns to scale). A smooth function with the two properties above would then be characterized by $\varepsilon'(\omega) > 0$ and $\varepsilon''(\omega) < 0$: the marginal value is positive but decreasing.

One of the simplest mathematical functions that can be adopted to express $\varepsilon(\omega)$ is the powered semi-logistic one:

$$\varepsilon(\omega) = \left(\frac{\alpha\omega}{1+\alpha\omega}\right)^{\beta} \omega \geq 0 \ \beta > 0 \quad (4)$$

By plugging (3) into (1), and solving for the MVW, a relationship linking MVW to WIC (ω) is obtained:

$$MVW = \omega^{-1}\left(\frac{\alpha\omega}{1+\alpha\omega}\right)^{\beta} \quad (5)$$

This allows us to infer the marginal value of water on the basis of the water intensity, once the values of a and β have been set. We calibrated the values for these parameters using some estimates by Moolman et al. (2006), who computed the MVW for five categories of fruits in South Africa, in the year 2002, and our own estimates of the industrial water intensity for the year 2004 (Roson and Damania 2017). The beta parameter is calibrated by imposing that MVW equals 1.312 (simple mathematical average of the estimates by Moolman et al. cit.) when WIC (ω) is 0.01039 (our estimated value for Vegetables and Fruits in South Africa). The alpha parameter is simultaneously obtained through numerical optimization, imposing the requirement that the variance of MVW values by Moolman et al. equals the variance of MVW across South-African industries (excluding the outlier Services). The computed values are 0.637 for alpha, 0.855 for beta. Table 4.8 presents the corresponding MVWs.

Notice that, the higher the *average* productivity of water (value of output per m^3, the inverse of the WIC), the higher the *marginal* value of water. In this respect, allocating water resources on the basis of the relative industrial water productivity (as it is done in the DRES scenario) is conceptually equivalent to allocating water on the basis of the relative marginal values.

The output elasticity of water is the percentage increase in gross production volumes obtained through higher water utilization. If no adjustment takes place in the production processes and in the use of other factors, then the elasticity is just the product of MVW and WIC. To get more meaningful effects when water availability is varied, we allow in this study some implicit adjustment in complementary factors, by expressing the output elasticity of water (η) as a linear function of the product (ε):

$$\eta = \gamma + \delta\varepsilon \tag{6}$$

where the γ and δ parameter values are set so that the average elasticity is 0.8 and the standard deviation is 0.2. Table 4.9 shows the elasticities obtained in this way.

Appendix 3: The Construction of Regionally Differentiated Impact Scenarios

Simulations under the DRES and DRUS scenarios are based on the assumptions that regions, in which potential demand for water exceeds sustainable supply, differ in their capability of absorbing the excess demand (water deficit). The absorption percentages applied in the various cases are based on a mixed qualitative-quantitative analysis of the relevant characteristics, where we keep distinct the potential of technological innovation from the degree of flexibility in the economic structure and trade flows.

Looking first at the innovation side, notice that a number of technologies and management options can be put in place to improve the water efficiency (lowering demand) and/or expanding the water supply. Theoretically, the different options could be ranked in terms of economic efficiency, from the lowest to the highest unit cost, and those whose unit cost (possibly including externalities) falls below the shadow value of water (increasing as the water gets scarcer) should be selected (WRG 2009). In practice, however, the technological response to the water stress is much more complicated, as a variety of factors (technical, political, institutional, safety, etc.) ultimately affects the choice among the different technology options (Becker et al. 2010).

Table 4.8 Industrial MVW 2004 (US$/M^3)

	N_America	C_America	S_America	W_Europe	E_Europe	MENA	Sahel	C_Africa	S_Africa	C_Asia	E_Asia	S_Asia	SE_Asia	Australasia
Rice	1.247	1.025	1.151	1.494	1.290	1.352	1.165	1.194	1.013	0.847	1.359	1.102	1.163	1.071
Wheat	1.237	1.531	1.344	1.513	1.197	1.375	1.113	1.847	1.449	1.006	1.282	1.224	1.275	1.190
Cereals	1.278	1.303	1.267	1.489	1.267	1.357	0.998	1.375	1.181	1.085	1.238	1.158	1.356	1.165
VegFruit	1.372	1.451	1.199	1.724	1.508	1.553	1.327	1.400	1.312	1.408	1.683	1.413	1.258	1.614
Oilseeds	1.495	1.570	1.582	1.830	1.299	1.413	1.064	1.519	1.236	0.528	1.423	1.309	1.915	1.231
Sugar	1.464	1.282	1.192	1.781	1.390	1.474	1.083	1.670	1.347	1.181	1.379	1.321	1.278	1.286
Oth Crops	1.519	1.490	1.512	2.149	1.468	1.708	1.280	1.403	1.505	1.634	2.096	1.927	1.498	1.568
Oth Agr.	1.912	1.730	1.653	2.303	1.768	1.699	1.181	1.353	1.600	1.498	1.858	1.503	1.436	1.932
Extraction	1.910	1.832	2.063	2.293	1.815	1.843	1.873	1.981	1.816	1.857	2.317	2.044	2.077	2.462
P.Food	3.098	3.425	3.037	3.628	2.442	3.067	2.917	2.917	2.917	2.917	3.436	2.969	3.098	3.650
Textiles	3.036	2.883	2.944	3.472	2.417	3.141	2.719	2.719	2.719	2.719	3.452	3.188	2.975	3.181
Light Man	3.825	3.426	3.343	4.264	2.914	3.158	3.076	3.076	3.076	3.076	3.396	2.655	3.271	4.113
Heavy M.	3.481	3.211	3.027	3.908	2.782	2.845	2.793	2.793	2.793	2.793	3.167	2.591	2.926	3.648
Utilities	2.126	2.040	1.739	2.370	2.059	1.922	1.871	1.871	1.871	1.871	2.243	1.999	1.916	2.293
Services	19.508	18.617	19.970	27.501	19.657	18.041	17.410	17.410	17.410	17.410	25.398	19.973	17.921	25.188

Table 4.9 Industrial output elasticity of water

	N_America	C_America	S_America	W_Europe	E_Europe	MENA	Sahel	C_Africa	S_Africa	C_Asia	E_Asia	S_Asia	SE_Asia	Australasia
Rice	0.9292	1.3853	1.0672	0.7654	0.8851	0.8360	1.0426	0.9968	1.4232	2.2836	0.8316	1.1668	1.0468	1.2464
Wheat	0.9397	0.7534	0.8415	0.7593	0.9933	0.8211	1.1435	0.7009	0.7833	1.4489	0.8925	0.9562	0.8995	1.0026
Cereals	0.8958	0.8737	0.9067	0.7672	0.9068	0.8325	1.4817	0.8215	1.0160	1.2093	0.9392	1.0540	0.8331	1.0428
VegFruit	0.8233	0.7822	0.9902	0.7140	0.7609	0.7474	0.8540	0.8070	0.8657	0.8028	0.7201	0.7998	0.9166	0.7326
Oilseeds	0.7653	0.7429	0.7398	0.7024	0.8768	0.8001	1.2655	0.7572	0.9419	1.6063	0.7950	0.8686	0.6958	0.9469
Sugar	0.7769	0.8922	0.9996	0.7072	0.8123	0.7730	1.2149	0.7221	0.8396	1.0168	0.8191	0.8589	0.8963	0.8886
Oth Crops	0.7572	0.7670	0.7595	0.6853	0.7752	0.7163	0.8939	0.8055	0.7617	0.7287	0.6870	0.6951	0.7641	0.7433
Oth Agr.	0.6961	0.7132	0.7251	0.6816	0.7087	0.7176	1.0160	0.8352	0.7357	0.7641	0.7000	0.7624	0.7887	0.6948
Extraction	0.6962	0.7022	0.6883	0.6818	0.7037	0.7012	0.6988	0.6920	0.7037	0.7001	0.6814	0.6890	0.6877	0.6793
P.Food	0.6757	0.6751	0.6758	0.6749	0.6795	0.6757	0.6762	0.6762	0.6762	0.6762	0.6751	0.6760	0.6757	0.6749
Textiles	0.6758	0.6763	0.6761	0.6750	0.6799	0.6756	0.6771	0.6771	0.6771	0.6771	0.6751	0.6755	0.6760	0.6755
Light Man	0.6748	0.6751	0.6752	0.6746	0.6762	0.6755	0.6757	0.6757	0.6757	0.6757	0.6751	0.6775	0.6753	0.6746
Heavy M.	0.6750	0.6754	0.6758	0.6747	0.6768	0.6765	0.6767	0.6767	0.6767	0.6767	0.6755	0.6780	0.6762	0.6749
Utilities	0.6860	0.6892	0.7122	0.6805	0.6884	0.6954	0.6990	0.6990	0.6990	0.6990	0.6829	0.6911	0.6958	0.6818
Services	0.0000	0.0000	0.0000	0.0000	0.0000	0.0000	0.0000	0.0000	0.0000	0.0000	0.0000	0.0000	0.0000	0.0000

Table 4.10 Regional technology potential index

MENA	Sahel	C_Asia	E_Asia	S_Asia
31	28	30	35	31

We therefore rely on a qualitative index of technology potential for each of the potentially water stressed macro-regions, based on a subjective evaluation of several options and characteristics. Because of the subjective and qualitative nature of this index, the latter should be interpreted as expressing an informed scenario, rather than as a solid scientific appraisal of (future) technical capability in the regions.

We consider three important classes of technology or management options:

1. Desalination
2. Enhanced irrigation techniques and reduced evaporation
3. Water reuse.

For each of them, we identify five "facilitating factors", possibly making the implementation of each option more likely:

1. Physical conditions (e.g., desalination projects will be more effective if most of the urban centres are found along the coast).
2. Factor availability (e.g., access to energy sources for desalination).
3. Institutional capacity (efficient level of government, quality of public institutions).
4. Human and physical capital (relevant for large and complex projects).
5. Demand potential (e.g., enhanced irrigation is primarily targeted to agriculture, therefore its effectiveness depends on the share of agricultural water on total water consumption).

We assign to each factor in each region and for all the three alternatives above a simple scoring system: 1 (poor), 2 (average), 3 (good). A "Technology Potential Index" (Table 4.10) is quite naturally obtained by simply adding up all the given points. The higher this index, the easier is the expected capability of a region to adjust to water deficits through the introduction of new technologies and more efficient management techniques.

A second adjustment mechanism is related to the endogenous changes in the regional economic structure. Indeed, when actual water availability turns out to be lower than what would be required for production and consumption purposes, the consumers' utility diminishes and the productivity in water-using industries declines. Even in the absence of a formal market for water resources, scarcity is transmitted as a price signal, and a structural adjustment takes place in the economic system, alleviating the overall impact of the negative shock for the economy. What is maybe less known is that the same process leads to an improvement in the aggregate water efficiency or productivity (water per unit of output), whose magnitude—however—depends on a series of specific characteristics of the economic system under consideration.

Table 4.11 Agricultural output change

Region	Var. (%)
MENA	−8.71
Sahel	−7.80
Central Asia	−12.20
East Asia	−4.22
South Asia	−5.15

Table 4.12 Decomposition of excess water demand absorption

	MENA	Sahel	C_Asia	E_Asia	S_Asia
Internal alloc.	0.42	0.44	0.36	0.52	0.5
Tech solutions	0.32	0.26	0.3	0.4	0.32
Water cuts	0.26	0.3	0.34	0.08	0.18

Many factors contribute in determining the structural flexibility, and it is not easy to ascertain what economies could respond better and why. To shed some light on this issue, we performed a simple numerical experiment with the global general equilibrium model. In each of the potentially water stressed macro-regions, we simulated a −10% reduction in multi-factor productivity in agriculture, which is the sector where most of the water is utilized. The consequent drop in total agricultural output volume is shown in Table 4.11.

A CGE model cannot capture all the factors and characteristics affecting the actual degree of flexibility in a certain economy. Nonetheless, a simple experiment like the one above can offer an order of magnitude, or at least can suggest a ranking of the regional economies from the most rigid one (Central Asia) to the most flexible one (East Asia), in terms of absorption of productivity shocks in agriculture, possibly induced by water scarcity.

We combine the ranking provided by Tables 4.10 and 4.11 to split the absorption of the excess water demand in the three components: internal structural adjustment, technical and management solutions, and reduction in water delivery. The latter component, which is obtained as a residual, determines the amount of decrease in water delivery (with effects on productivity) in the scenarios DRUS and DRES (Table 4.12).

References

Alcamo J, Döll P, Henrichs T, Kaspar F, Lehner B, Rösch T, Siebert S (2003) Development and testing of the WaterGAP 2 global model of water use and availability. Hydro Sci J 48:317–337

Arnell NW (2004) Climate change and global water resources: SRES emissions and socio-economic scenarios. Glob Env Change 14:31–52

Becker N, Lavee D, Katz D (2010) Desalination and alternative water-shortage mitigation options in Israel: a comparative cost analysis. J Water Resour Prot 2:1042–1056
Deikman J, Petracek M, Heard JE (2012) Drought tolerance through biotechnology: improving translation from the laboratory to farmers' fields. Curr Opin Biotech 23:243–250
Hertel T, Tsigas M (1997) Structure of GTAP. In: Hertel T (ed) Global trade analysis: modeling and applications. Cambridge University Press, Cambridge
Kriegler E, O'Neill BC, Hallegatte S, Kram T, Lempert RJ, Moss RH, Wilbanks T (2012) The need for and use of socio-economic scenarios for climate change analysis: a new approach based on shared socio-economic pathways. Glob Env Change 22:807–822
Moolman CE, Blignaut JN, van Eyden R (2006) Modelling the marginal revenue of water in selected agricultural commodities: a panel data approach. Agrekon 45:78–88
O'Neill BC, Kriegler E, Riahi K, Ebi KL, Hallegatte S, Carter TR, Mathur R, van Vuuren DP (2014) A new scenario framework for climate change research: the concept of shared socio-economic pathways. Clim Change 122:387–400
Reimer JJ (2012) On the economics of virtual water trade. Ecol Econ 75:135–139
Rosegrant MW, Cai X, Cline SA (2002) Global water outlook to 2025. In: Averting an impending crisis. IFPRI, Washington DC and IWMI, Colombo, Sri Lanka
Roson R (2016) Simulating the macroeconomic impact of future water scarcity. In: Water global practice discussion papers, The World Bank, Washington DC
Roson R, Damania R (2017) Simulating the macroeconomic impact of future water scarcity: an assessment of alternative scenarios. J Policy Model 39:1141–1162
Ward FA, Michelsen A (2002) The economic value of water in agriculture: concepts and policy applications. Water Policy 4:423–446
Wilson C (2018) Biodegradable technology brings 'higher cotton yield at less cost to the environment'. http://www.abc.net.au/news/2018-09-08/water-efficient-growing-technology-brings-higher-cotton-yield/10195578. Accessed 10 Sept 2018
WRG, The 2030 Water Resources Group (2009) Charting our water future: economic frameworks to inform decision-making. https://europa.eu/capacity4dev/euwi-community-space/document/charting-our-water-future-economic-frameworks-inform-decision-making. Accessed 8 Sept 2018

Prof. Roberto Roson received a doctoral degree in Economics from the University of Umeå (Sweden). He has been "visiting fellow" at the Free University of Amsterdam, at the University of Warwick, at S. Francisco Xavier U. (Sucre, Bolivia), San Simon U. (Cochabamba, Bolivia), University of Barcelona, as well as at the World Bank (Washington) and at the Institute of Developing Economies—JETRO (Chiba, Japan). He is currently Professor at Ca'Foscari University, Venice. He is also affiliated with the GREEN Research Centre of Bocconi University in Milan, as well with the Department of Economics of Loyola Andalusia University in Seville, Spain. He is the author of several articles published in books and international scientific journals, and coordinator of several applied research projects, including the European FP7 project WASSERMed. He has been consultant for several organizations, including: European Commission (JRC), World Bank, United Nations and FAO. Roson in Scientific Director of the "CF Applied Economics" Research Centre for Applied Economic Research. His research interests deal primarily with environmental economics, computable models for economic policy simulation and industrial organization of service industries.

Chapter 5
TERM-H2O Modeling of Droughts in Australia and California

Glyn Wittwer

Abstract In the early years of implementation of Australia's 2007 Water Act, analysis with TERM-H2O put water policy into perspective. Drought resulted in economic hardship, whereas legislated water buybacks provided farmers with an adjustment option. TERM-H2O predicted with reasonable accuracy the impact of abnormally dry or wet years on irrigation water prices within the Murray-Darling Basin. The theory of TERM-H2O was introduced to a version of USAGE-TERM modified to depict the impact of drought on California. The unsustainable use of ground-water in response to the Californian drought indicates a need for water policy reforms in the state.

Keywords Drought impacts · Water reforms · Farm factor mobility · Water trading

5.1 Background

One objective of this chapter is to explain the role that CGE modeling played in policy analysis of water reforms in Australia. In the circumstance of an adverse event such as drought, the blame for economic losses arising from drought will be assigned by critics of a coincident policy, in this case the 2007 Water Act (Australian Government 2007), to that policy. For CGE modelers, the analogy is that of tariff cuts: a global recession in the early 1990s led to substantial job losses. Many attributed recession-induced job losses to tariff cuts rather than background economic circumstances. A model helps us to analyze the marginal impacts of a policy, disentangled from changes induced by non-policy events.

This chapter also outlines the preparation of a CGE model of the irrigation regions of California, based on TERM-H2O, first used to model impacts within the Murray-Darling Basin. Since California is more reliant on groundwater for irrigation than

G. Wittwer (✉)
Center of Policy Studies, Victoria University, PO Box 14428, Melbourne 8001, Australia
e-mail: glyn.wittwer@vu.edu.au

G. Wittwer (ed.), *Economy-Wide Modeling of Water at Regional and Global Scales*, Advances in Applied General Equilibrium Modeling,
https://doi.org/10.1007/978-981-13-6101-2_5

the Murray-Darling Basin, the response to drought has differed markedly from that in the basin. However, there is considerable overlap in policy implications.

5.2 The Context for Model Modifications

5.2.1 Observed Changes in Water Usage from Year to Year

The need to make substantial modifications to the supply-side of a CGE model that includes water became apparent from water usage data prepared by the Australian Bureau of Statistics. In the drought of 2002, water use in rice production dropped by more than 70% relative to the previous year. In 2008, water used in rice production was only 2% of that in 2006 (ABS 2009a). The challenge was to approximate observed changes in water usage and in water prices in CGE modelling of drought.

Table 5.1 shows the substantial variation in rainfall and runoff across the Murray-Darling Basin from 2007–08 to 2016–17. The data shown in the table includes updates from earlier studies (Dixon et al. 2011; Wittwer and Griffith 2011), including the relatively wet years of 2010–11 and 2011–12. The additional years of data serve to illustrate features of TERM-H2O that are useful in both dry and wet years.

2008–09 was the last season of the millennium drought in the southern basin; in the northern basin, the drought broke early in 2008. The water price shown reflects temporary water trades in the southern basin.[1] The water price reflects current water availability, rainfall conditions and farm output prices. For example, in 2007–08 prior to the GFC, a biofuels boom drove up crop prices and the marginal product of water with it. Table 5.1 (row 13) shows that the water price in 2008–09 was $298/ML with drought still prevailing in the southern basin, whereas in the previous year it was $680/ML prior to the GFC-induced collapse in commodity prices. Rice is the most water-intensive of crops grown in the basin and therefore the most responsive to changes in water price. In the past decade, the volume of water used in rice production has varied from 27 GL in 2007–08 to 1434 GL in 2012–13 (Table 5.1, row 3).

Table 5.1 has embedded in it other aspects of on-farm water management. A La Nina event resulted in two years of record rain surpluses in a substantial part of the basin for the period ending June 2012.[2] This replenished dams under the jurisdiction of irrigation authorities, filled creeks and dams from which farmers extracted water (see Table 5.1, row 12) and also rejuvenated groundwater supplies. An amendment to the Water Act in 2011 introduced carryover provisions for annual water allocations: these enable irrigators to make on-farm allocation decisions independent of irrigation

---

[1]Temporary water refers to a volume of water allocated in a single season. Permanent trades involve an exchange of a water right.

[2]See http://www.bom.gov.au/jsp/awap/rain/archive.jsp?colour=colour&map=decile&year=2012&month=6&period=24month&area=md. Accessed 29 August 2016.

Table 5.1 Water use in the Murray-Darling Basin

Year	2007–08	2008–09	2009–10	2010–11	2011–12	2012–13	2013–14	2014–15	2015–16	2016–17
1. Area irrigated ha*10^6	1.0	0.9	1.0	1.2	1.4	1.6	1.6	1.4	1.2	1.4
2. Not irrigated ha*10^6	94.6	95.1	94.2	83.8	93.6	88.4	89.1	87.8	82.8	88.7
Crop	*Water used (GL)*									
3. Rice	27	101	205	755	1134	1434	912	876	299	935
4. Cotton	283	793	764	1789	1906	2735	2676	1114	1294	2420
5. Other annuals	1044	1066	719	537	725	1334	1368	1226	815	912
6. Grapes	434	439	428	303	365	463	415	431	428	374
7. Other perennials	356	374	450	379	475	567	713	502	664	598
8. Pasture, hay, silage	997	719	998	744	1270	2041	1941	2025	1438	1423
9. Total	3141	3492	3564	4507	5875	8574	8025	6174	4938	6662
	Source (GL)									
10. Irrigation schemes	NA	1573	1830	1705	2768	4228	3494	3212	2088	2874
11. Groundwater	NA	1069	989	570	568	686	863	844	926	713
12. Other (dams, creeks, etc.)	NA	850	745	2232	2539	3660	3668	2118	1924	3075
13. Water price (AUS$/ML)[a]	680[a]	298[a]	148[a]	26[a]	16[a]	51	64	106	168	85

Source ABS 2009a, various years

[a]http://www.murrayirrigation.com.au/water/water-trade/water-exchange-history/ provided Murray Irrigation weighted sales that were reasonable indicators of average weighted southern MDB trading prices

authorities.[3] Following the wet two year period, some water allocations from those years were used in the three years from 2012–13 to 2014–15.

Another feature of Table 5.1 is that although groundwater extractions increase when water becomes scarce, they are not sufficient to counter shortfalls in water allocations. For example, in 2008–09 when the southern basin was still in drought, groundwater extractions amounted to 1069 GL. In 2012–13, when total water used was 5000 GL more than in 2008–09, groundwater extractions were only 383 GL less (Table 5.1, rows 9 & 11). That is, changes in groundwater extractions fall far short of offsetting fluctuations in volumes diverted from other irrigation sources.

Turning to perennials, water used for grapes dropped below 400 GL only in the La Nina years 2010–11 and 2011–12, when rainfall was sufficiently high to reduce irrigation requirements (Table 5.1, row 6). The upward trend in water used by other perennials reflects a substantial increase in almond plantings since 2007.[4]

Table 5.1 also shows water used in pasture, hay and silage (row 8). These feed products are either inputs into livestock production locally, or are sold outside the basin. The table shows a 40-fold difference in the price of water between 2007–08 (i.e., \$680/ML) and 2011–12 (\$16/ML) (Table 5.1, row 13). Even in drought when the price of livestock feed rises, it is highly probable that a livestock farmer within the basin would gain by selling water to fund feed purchases instead of using scarce water to grow on-farm livestock feed.

5.2.2 *Devising Appropriate Model Theory from Observed Data*

Table 5.1 points us to the necessary features of our CGE model of irrigation activity, TERM-H2O. First, we require a split between dry-land and irrigated agriculture. The table (lines 1 & 2) indicates that irrigation makes up less than 2% of the land used in agriculture in the basin. ABS (2008) estimate that around 30% of agricultural output in the basin arises from partly or wholly irrigated activity, and exclusively dry-land agriculture accounts for the remaining 70%. Lines 1 & 2 of the table show that the area of dry-land and irrigated agriculture varies from year to year, depending on water availability and output market conditions. This indicates that the theory of our model requires some mobility of irrigable land between irrigated and dry-land technologies.

As mentioned above, line 6 of Table 5.1 indicates that rainfall variability alters irrigation water requirements. Water availability and rainfall are exogenous in the CGE model, but we need to distinguish between the two. That is, if there is a rainfall deficit, we need to shock water supply in the model, which will lead to substitution away from water in the production function of the irrigated industry. We treat

[3] See and http://www.mdba.gov.au/about-us/governance/water-act. Accessed 6 September 2016.

[4] See http://growing.australianalmonds.com.au/wp-content/uploads/sites/17/2014/06/Australian-Almond-Insights-2013-14-LR-WEB.pdf. Accessed 29 August 2016.

impacts in dry-land agriculture differently, by ascribing production shocks to depict the impact of drought.

Wide variation from year to year in water usage for rice and cotton indicate that we require flexibility in allocation of farm factors for annuals. Line 5 of Table 5.1 refers to other annuals that are less water-intensive, particularly vegetables. Since water is a smaller share of the total costs of production for other annuals, the responsiveness to changes in the water price is smaller than for rice or cotton. As was the case for grapes, other annuals used less water than usual in 2010–11 and 2011–12 due to above average rainfall. Cotton is split into two technologies, dry-land and irrigated, since it can be grown without irrigation in wet seasonal conditions.

Given the substantial investments that go into establishing vineyards and orchards, we need to depict factor rigidity in perennials. When water is scarce, as was evident in 2007–08 and 2008–09, perennial sectors purchase water from other users in response to diminished water allocations. Even if the water price soars, the costs of destruction to perennial plantations in terms of foregone future income may far exceed the additional water costs in a water-scarce year. Factor rigidity is imposed by including specific capital for perennials, whereas annuals use capital that is mobile between different farm activities.

Specific capital (i.e., the herd) in also used in livestock production. TERM-H2O includes substitutability between irrigated land (defined in Sect. 3) and feed inputs in livestock production. We can model a similar response by including hay & forage as a substantial input to livestock production. In turn, hay & forage is split into dry-land and irrigated technologies. The specific capital in livestock production ensures that the producer response to worsening water scarcity is either to substitute from irrigated land to feed inputs in the case of TERM-H2O or, as in the modified version of USAGE-TERM outlined in Sect. 7, from irrigated feed inputs on-farm to feed purchased from elsewhere.

Changes in relative output prices alter the allocations of mobile capital, operator labor, dry land and irrigable land between activities. These factors follow a constant elasticity of transformation (CET) form.

Another important feature of TERM-H2O concerns water trading possibilities. The main stylized assumption is that irrigation water is perfectly tradable between irrigation sectors and regions of the southern basin. That is, water is traded at a single price in the southern basin in the model, which has approximated reality. In the northern part of the Murray-Darling Basin, which consists of far-flung tributaries, we assume that water is tradable within a region but not between regions. In the original version of TERM-H2O, a single region depicted the northern basin. This was acceptable if scenarios concentrated on the southern basin and results for the northern basin were not reported, as in Dixon et al. (2011) and Wittwer and Griffith (2011). In subsequent projects undertaken for clients, the northern basin was of interest. In preparation for these projects, TERM-H2O was split into catchment regions in the northern basin, with no water trading between regions.

5.3 Main Modifications in TERM-H2O

Figure 5.1 shows us that TERM-H2O's Leontief bundle of intermediate input and primary factor follows the conventional theory of a CGE model in the Dixon et al. (1982) school. The key difference from a conventional dynamic CGE model emerges in the elaboration of farm factor mobility. The primary factor in Fig. 5.1 is a CES nest of hired labor, general purpose capital and land and operator. Hired labor is a CES nest of occupations. General purpose capital, including farm implements and sheds, is mobile at the farm level between different activities. The land and operator factor is a CES bundle of specific capital, operator labor and land. Annual crops will use a negligible amount of specific capital, whereas it is significant in livestock and perennial production.

As is evident at the bottom of Fig. 5.1, land is a CES nest of three types of agricultural land, namely dry land, irrigable land without water and irrigated land. If water allocations fall, some irrigated land will switch to irrigable land without water, and may switch back in subsequent time periods with the restoration of usual water allocations. In terms of hectares, dry land, which cannot be irrigated, dominates farming in the Murray-Darling Basin (Table 5.1, rows 1 & 2). Without water, irrigable land is far less productive per hectare than with water. With water using technologies constant, a fixed amount of water inclusive of effective rainfall is required per hectare of irrigable land for a given output.

As mentioned earlier, the original TERM-H2O elaborated in Dixon et al. (2011) includes substitutability between land and feed inputs in livestock production, though this is not shown in Fig. 5.1. The subsequent USAGE-TERM version of the model includes a larger land input in feed production and a larger feed input into livestock production than the initial TERM-H2O representation.

TERM-H2O includes a dry-land technology and an irrigated technology for each of the following: cereal, dairy cattle, other livestock, cotton, fruit and other agriculture. Rice, grapes and vegetables were assigned irrigation technologies only.

5.4 Policy Corrections

It is one thing to invest much effort in devising a CGE model for a task such as analyzing irrigation agriculture. But is it important to depict the observed water usage variability between users and implied factor mobility that is shown in Table 5.1 within a CGE model? The Australian experience is that such a model is crucial in water policy analysis. Without it, exaggerated claims concerning policy measures may prevail without challenge because such claims arise in part from the assumption that farm inputs are rigid.

The 2007 Water Act was legislated late in the Howard government era in an attempt to address over-allocations of irrigation water within the basin. The formation of the Murray Darling Basin Authority (MDBA) proceeded under the Act.

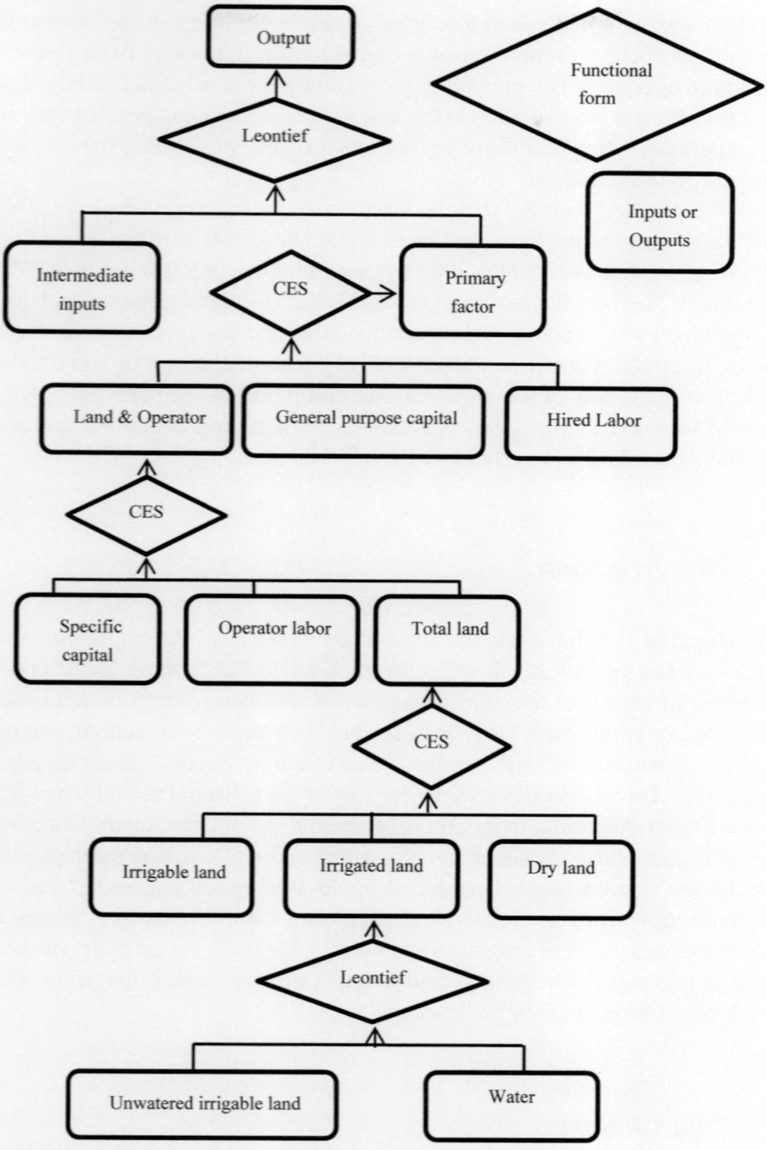

Fig. 5.1 Production function for farm industries

The MDBA was assigned a complex array of responsibilities. In the context of this chapter, it is sufficient to mention two, namely water buybacks from farmers and infrastructure upgrades. The prevailing view among economists is that buybacks are a more cost efficient way of addressing sustainable water management than infrastructure upgrades. The latter dominate the costs of the Act, which over its lifetime may exceed AUS$12 billion.

The Act coincided with the middle of a three-year period (2005–06 to 2008–09) in which large areas of the basin and its mountain headlands suffered record rainfall deficits. Water buybacks under the 2007 Act were blamed for job losses in the Murray-Darling Basin. The culprit was drought: TERM-H2O modeling showed that drought-induced job losses were many-fold greater than the job impacts from water buybacks (Dixon et al. 2011; Wittwer and Griffith 2011). The restoration of some jobs with the end of drought was observable in basin communities. In particular, rice mills mothballed for around three years opened again when the millennium drought was superseded by the La Nina event of 2010 to 2012.

5.4.1 Why Irrigation?

Should irrigation ever have taken started in the first place? Australia established a number of soldier settlement schemes after each world war. These were ill-conceived for a number of reasons. First, they assumed that returned soldiers would make able farmers. For some this may have been so, but for others, such schemes may have resulted in a mismatch of skills, aptitude and vocation. Second, using an argument propounded by Davidson (1969) in the context of the subsequent Ord River Scheme in far north-west Australia, irrigation schemes did not reflect Australia's comparative advantage as a land-abundant, people-scarce nation. Davidson's perspective was that the nation's endowments best suited broad-acre agriculture rather than irrigation technologies which are relatively capital- and labor-intensive. A legacy of the unsound economic basis of irrigation schemes in Australia is that many of the products of irrigated regions were subjected to much greater market distortions than the main outputs of broad-acre agriculture.

5.4.2 Policy Reforms

Clearly, a hypothetical question such as whether Australia would have been better off had irrigation schemes not been established in the Murray-Darling Basin is of little policy relevance. More relevant is policy formation that aims to improve economic, social and environmental outcomes, given that irrigation activity is a part of the basin's present economic structure.

Also of relevance are models that will enable us to assess the benefits of proposed irrigation schemes. Wittwer and Banarjee (2015) analyzed the impacts of a possible scheme in far north-west Queensland using a different version of TERM. They found that in all probability, the scheme would not yield net economic benefits. Despite evidence that the original Ord River Scheme in northern Western Australia was not ever worthwhile, the Federal and Western Australian governments announced funding for an expansion of the scheme in 2009.[5] The questionable viability of the extension has been worsened by cost blowouts (Kagi 2016).

The environmental health of the Murray-Darling Basin has been an ongoing concern for communities, environmentalists, economists, farmers and politicians for many decades. Water policy reforms in the basin have arisen from a long and painful process. The Council of Australian Governments made substantial strides towards water reforms several decades ago (COAG 1994). A cornerstone of the reforms was to separate agricultural land ownership from water ownership. This was an important and necessary step in the use an instrument with the potential to improve both environmental and economic outcomes: an appropriate price for water.

5.5 Myth-busting with Available Data: Buybacks and Economic Activity in the Basin

This section provides an example of how the task of preparing a modified CGE model and database for scenarios in the Murray-Darling Basin contributed to policy debate. If our model preparation is thorough, we may improve our understanding of the institutions and issues at play in the economy. In the context of the basin, under the 2007 Water Act, between 27 and 37% of surface water used in irrigation in the basin was to be purchased by the Commonwealth from farmers. The process aimed to be voluntary, with full compensation to farmers at market prices for water.

The following lists myths that to the present day are propounded by local councilors, and state and Federal politicians including cabinet ministers.

Myth 1: The Murray-Darling Basin agricultural base consists mainly of irrigated farming.

ABS (2008) estimated that in 2005–06, around 30% of the Murray-Darling's output arose from irrigated agriculture and 70% from dry-land farming. As a share of basin-wide income of all sectors (agricultural plus non-agricultural), irrigated agriculture's share is 5% and dry-land agriculture's share 11% (based on ABS cat. 7121.0, ABS census data and ABS national accounts).

Myth 2: If some water is removed from irrigation uses, farming will collapse in the basin.

ABS Water Accounts indicate that in response to worsening water scarcity, dairy farmers in the basin reduced irrigation water usage by almost 65% in 2007–08 relative

[5] See http://www.lakeargyle.com/explore-and-learn/ord-river-irrigation-scheme/. Accessed 26 September 2018.

Table 5.2 Estimates of direct impacts of drought and buybacks on Murray-Darling Basin farming

	Drought 2007–08 relative to base[a]	Fully implemented buybacks (3500 GL) relative to forecast
Dry-land productivity	−20%	0
Irrigation: rain	−37%	0
Irrigation: water	−37%	−32%
Compensation	No	Full
Process	Involuntary	Voluntary

[a]Scaled to basin-wide impacts from estimates of southern basin impacts reported in Wittwer (2011)

to 2005–06—with a much smaller percentage loss in output (see Table 5.3). Strong dairy prices encouraged farmers to purchase fodder to compensate for shortfalls in irrigation water. Dairy farmers were able to funded fodder inputs by selling off their (diminished) water allocations at many-fold the water price of an average year. Farm factor mobility plays a key role in adjusting to short-term water deficiencies.

Myth 3: Buybacks result in job losses, as is evident in employment trends in the basin shortly after buybacks started.

Job losses in the basin arose out of the three year drought that affected the southern part of the basin between 2006–07 and 2008–09. Even if farmers were not compensated for water removed from irrigation activities, drought would still be unambiguously worse than buybacks, as shown in Table 5.2. The fully implemented buyback volume used in the table is 3500 GL.[6]

Dry-land activities account for at least two-thirds of basin farm output. Irrigation farming relies in part on rainfall, which is unaffected by buybacks. Supposing rainfall accounts for 20% and irrigation water 80% of the water used in irrigation farming. We can infer from Table 5.2 that at worst, farm output in the basin would decrease by about 8% ($= 1/3 \times 0.8 \times -32$) with 3500 GL removed from production. But since farm factors are mobile, the final output will be smaller than this. TERM-H2O modelling indicates that for every lost dollar of irrigation output, farm factor movements (most notably in livestock production) result in an increase in dry-land production of about half a dollar (Dixon et al. 2011). In addition, there is substantial movement of water away from crops with a relatively low average product of water (rice in particular).

Myth 4: Removing 3500 GL of water in the basin from production is equivalent to a permanent drought.

[6]The total water entitlement in the basin including environmental water is around 10,800 GL. All users including environmental water holders receive less than their full allocation in drier seasons. In 2012–13, irrigators used 8574 GL (Table 5.1, row 9) and environmental water holders around 2000 GL (from https://www.environment.gov.au/system/files/resources/187e1491-d72c-41ab-9e82-9528915e1c96/files/mdb-environmental-water-holders-report-2013.pdf (accessed 2 September 2016).

Table 5.2 shows that fully implemented buybacks will have a substantially smaller impact on farm output than drought. The data in Table 5.1, which includes years following the heightened debate which occurred around 2010, shows that "half implemented" buybacks have had a minor impact on water scarcity.

Myth 5: Infrastructure upgrades are more acceptable to basin communities than buybacks and have the potential provide environmental gains without economic hardship.

While the buyback process was still active, there was no shortage of sellers. Farmers have various motivations for selling part or all of their water entitlement. Such sales may contribute to debt repayments, farm restructuring and semi-retirement or retirement plans (Wheeler et al. 2014). In some instances, water title sales may help farmers make a dignified exit from the industry.

Buybacks continued despite the fiery community meetings over which the MDBA presided in October 2010 (Delaney 2010). Community anger may have arisen from a perception that farmers were a scapegoat for environmental problems within the basin. Community stress arose from the aftermath of a prolonged drought. Ongoing structural change meant that even without drought, such communities were vulnerable, with a shrinking workforce over time. Though the drought had ended, farmers faced new difficulties with the soaring Australian dollar in the wake of the mining boom. For grape growers within the basin, for example, Australia's falling competitiveness in the international wine market drove grape prices below the costs of production. The buyback process became a flashpoint for community anger amid the ongoing difficulties being faced within the basin.

There was an economic basis for the 2007 Water Act to include provision for irrigation infrastructure upgrades, to the extent that such infrastructure has public good attributes. From a qualitative economic perspective, there may have been some "low hanging fruit" among possible infrastructure upgrades, with a satisfactory outcome for the money outlaid. But the available evidence is that even if we accept reduced leakages, which return to the hydrological cycle, as part of the gains from infrastructure upgrades, the costs of additional water are several-fold the costs of buybacks (Adamson and Loch 2014).

If the policy objective of infrastructure upgrades is to maintain economic activity within basin communities, the upgrades are not an appropriate policy instrument. Wittwer and Dixon (2013) showed that with optimistic assumptions concerning gains in water efficiency arising from infrastructure upgrades, additional spending on essential services would have created four times as many jobs as upgrades in basin communities for each dollar spent.

5.5.1 The Myths Persist

In response to structural stress, we can expect mythmaking to continue in basin communities. As the population of south-eastern Australia continues to grow, relatively static population levels in many basin communities may be accompanied by rising median ages, a consequent reduction in workforce participation and a diminution in the share of national economic activity. For local governments, this is a concern. The gravest of consequences associated with rural stress arise from mental health in rural communities, and disturbing indictors such as suicide rates. These rates tend to spike during drought.

With each drought, concerns about water buybacks resurface. An El Nino event in 2015 resulted in moderate drought, worsening basin water scarcity and elevating the price of water. According to calculations from the rice processor Sunrice, the margin realized by rice growers on water for the 2016 harvest (i.e., the average product of water) after subtracting non-water variable costs was $278 per megalitre.[7] Some water was traded for more than $280 per megalitre in the basin late in 2015. At somewhat lower water trading prices, some rice farmers would reason that they could obtain a better return, after imposing a non-zero value on their own labor, by trading part or all of their water allocation and reducing or abandoning rice cultivation for the year.

In anticipation of local rice output falling by two-thirds relative to the previous year, rice processors announced job cuts in January 2016. Local mayors in the rice-growing heart of the basin were calling for environmental water to be made available for irrigation in response (Shields 2016). But had environmental water had been made available for irrigation, it would have allocated to more valuable uses for water than rice production during drought, as is evident from examining Table 5.1, rows 3 and 13.

5.6 Key Findings of Buyback Modeling

Dixon et al. (2011) used dynamic TERM-H2O to model the regional impacts of the Commonwealth buying 23% of water entitlements over a 10 year period in the southern basin. One key result was that for each dollar of lost irrigation output, dry-land output increased by around half a dollar as farm factors moved into dry-land production. Job losses in the basin were relatively minor.

Some groups including the NSW Irrigators' Council did not believe the modeled result that southern basin farm output would fall by only 1.3% with fully implemented buybacks (NSWIC 2010). Their argument was that with the rise in the price of water, producers would lose competitiveness in international markets. But in addition to modelling an increase in the price of water of around $100/ML, TERM-H2O results also indicated a fall in the rental on irrigated land which largely offset the water price increase, so that the impact on total costs of production was minor.

The most surprising result of all, and one that could only come out of a CGE model, was that aggregate consumption increased in the basin relative to forecast with fully

[7] See https://www.sunrice.com.au/corporate/newsroom/media-releases/2016-riverina-rice-harvest-stronger-yields-results/ and https://www.watertradingaustralia.com.au/past_sales.php. Accessed 31 August 2016.

implemented buybacks. Why was this, given that there was a reduction in farm output and regional GDP relative to forecast? It turns out that since farmers are net sellers (i.e., exporters) of water in the scenario, water sales contribute to farmers' terms of trade. By taking water out of production, the Commonwealth's purchases raise the price of water for entitlements. An appendix to Dixon et al. (2011) calculated the asset value of entitlements based on future expected annual water prices, inclusive of the expectation of occasional future droughts. The annuities from the payment to farmers for buybacks sales shifted out the consumption function in each region. The terms-of-trade gain meant that the increase in spending power outweighed the decrease in farm output in the basin.

5.7 Drought Modeling in Perspective

It is helpful for a modeler to check whether the results generated for a particular scenario are believable. Back-of-the-envelope checks use a combination of shocks, database weights and simplified theory to verify results. Dixon et al. (2011) presented their checks of buyback modeling in great detail.

At the time of the buyback study, drought modeling was another important way of checking and calibrating the model. The contents of Table 5.1 include years of observation that occurred after TERM-H2O was constructed and after initial simulations were undertaken. Therefore, the table provides a retrospective view of drought and buyback impacts. Wittwer and Griffith (2011) modeled the impacts of the drought from 2006–07 to 2008–09 in the southern basin.

The large ascribed supply shifts, in the form of dry-land productivity deterioration, and reduced water allocations and rainfall in irrigation industries, result in a modeling difficulty. In any farm sector that does not have relatively elastic demands, deterioration in productivity or decreases in water availability may result in implausibly large output price increases. These increases may be so large that revenue increases, via the price hike exceeding the reduction in output. In exceptional cases, such as California's almond production, this may be possible. In most cases, it is an unrealistic depiction of drought.

One consequence of the prolonged drought was mothballing of rice mills for three years. To bring this response to the modeled scenario, Wittwer and Griffith (2011, 2012) introduced excess capacity to TERM-H2O, following Dixon and Rimmer (2010). A sticky rental adjustment assumption in downstream food processing sectors reduces capital utilization and thereby reduces demand for drought-affected inputs. In the drought scenario, this lessened upward price pressure on farm outputs. It also brought greater realism to the model by reducing employment in processing sectors.

The estimated impact on the southern basin's employment level was the loss of 6000 jobs during the drought. Given the lost years of investment arising from the drought, 1500 jobs were lost in the drought simulation relative to forecast a number of years after full recovery from drought. Since water buybacks commenced during drought, jobs were being lost in the basin and wrongly attributed to buybacks.

Table 5.3 Comparing MDB modeled outcomes to observed outcomes

	Modelled outcome deviation from 2007–08 base (%)			Observed 2007–08 relative to 2005–06[b] (%)		
	Output	Price	Water used[a]	Output	Price	Water used[b]
	(1)	(2)	(3)	(4)	(5)	(6)
Cereal	−55.3	43.6	−78.8	−45.8	92.1	−9.9
Rice	−84.9	86.2	−90.7	−98.2	46.3	−97.8
DairyCattle	−13.6	29.5	−40.9	−26.5	52.0	−64.4
OthLivestock	−23.1	41.4	−44.6	−1.2	−9.2	−70.6
Grapes	−17.9	18.0	−49.0	2.7	44.6	−15.7
Fruit	−7.7	13.5	−23.1	9.3	7.3	−13.7
Vegetables	3.5	6.8	−1.4	21.8	14.9	−18.4

[a]Entire Murray-Darling basin
[b]Water used in irrigation production
Source Wittwer and Griffith (2012), Table 7.3

One way of comparing modeled results with observation is to report the marginal impacts of drought with the observed changes between 2005–06 and 2007–08. This is not quite comparing like with like, because drought was not the only contributor to changes in the two years to 2007–08. A biofuels boom occurred at the same time, which drove up cereal prices prior to the GFC. That had the effect of drawing farm factors including water towards cereals. The observed price for cereals jumped by 92% in this time, compared with the modeled impact, which did not account for the demand shock provided by the biofuels boom. The observed decrease in water allocated to cereal production was much smaller than the modeled reduction. The observed price of grapes also rose more than the modeled price. Consequently, the observed reductions in water use for cereals and grapes were smaller than the modeled reductions.

Overall, TERM-H2O approximated observed impacts reasonably. The observed water price hike from 2005–06 to 2007–08 exceeded \$600/ML. The modeled increase was \$285/ML. This was closer to the observed hike in 2008–09 relative to 2005–06 (a hike of \$254/ML, Murray Irrigation). The latter year followed the GFC, which ended the biofuels boom and thereby reduced demand for water.

A key finding from the scenario was that water trading played a substantial role in the response to drought. The average water price in 2007–08 (\$680/ML, Table 5.1, row 13) far exceeded the average product of water in rice production, so that the decline in rice production to 2% of the level of 2005–06 was unsurprising. In the model, rice output fell by 85% rather than the observed 98% (Table 5.3).

In summary, TERM-H2O modeling of buyback impacts and drought impacts provided reasonable approximations of each. Table 5.1 shows that drought drove up the water price many-fold, and wet years brought it down. As indicated in footnote 6, environmental water purchases amounted to around 2000 GL by 2012–13. Though this is little more than half the 3500 GL mentioned in this chapter, it is sufficient to observe whether a reduction in entitlements has a dramatic impact on water scarcity. Water prices shown in row 13 of the table suggest that moderate reductions in water availability have only a modest impact. Seasonal rainfall conditions, water allocations, expectations and commodity prices all contribute to the price of water.

5.8 Modifying USAGE-TERM to Depict Californian Drought

USAGE-TERM is a multi-regional CGE model of the US economy in the TERM stable (Wittwer 2017). The state-based master database of the model was modified to include 12 counties in and near California's Central Valley as separate bottom-up CGE regions (Fig. 5.2). These counties are Butte, Colusa, Fresno, Glenn, Merced, Kern, Kings, Madera, San Joaquin, Stanislaus, Tulare and Yolo. The modified theory of USAGE-TERM reflects Fig. 5.1. Wittwer (2015) presents more detail on these modifications.

In the original version of USAGE-TERM, the representation of farm sectors in the BEA input-output table was amended (see http://www.bea.gov/industry/io_annual.htm). Instead of representing agricultural industries by farm type, they are represented

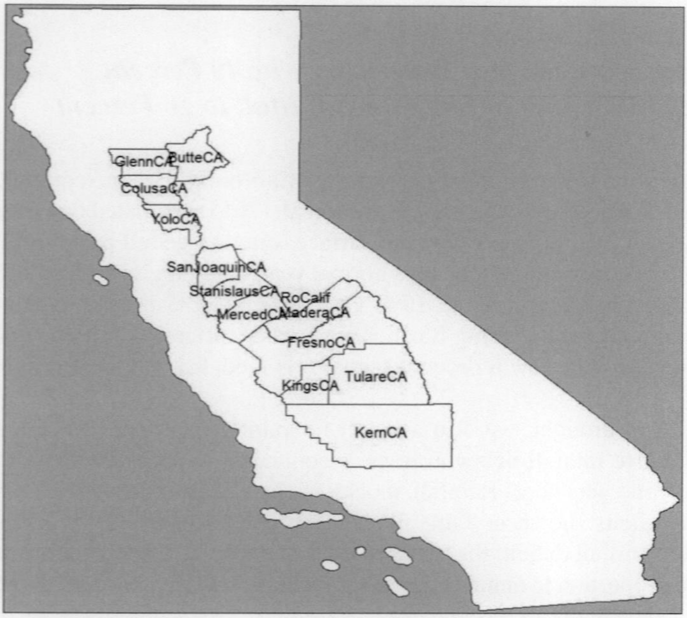

Fig. 5.2 Bottom-up Californian regions in the water version of USAGE-TERM

by outputs. This implies that on-farm factors may move between crop types as market conditions alter, enabling a given farm type to produce several outputs: the CET functional form provides factor mobility.

Concerning irrigation water, the modified version of the model includes satellite water accounts for irrigation sectors. These are based on USDA agricultural census data. Such data include acres of plantings and crop outputs. We can impose varying degrees of water tradability within the model. With one closure (that is, choice of endogenous and exogenous variables), water is tradable between all irrigators within a county, but with the exception of Kings and Kern counties, not tradable between counties. To assume that water is tradable within a county is reasonable on engineering grounds, but not on institutional grounds. We can alter the closure to prevent trading between certain types of irrigation activity, thereby mimicking institutional constraints to some extent.

We impose an arbitrarily low initial unit value on water (i.e., \$50 per acre-foot).[8] Our interest is in how much water prices increase as scarcity worsens. Since the average product of water in production of some crops is only a few hundred dollars per acre-foot, rising water scarcity will induce producers of crops with low average products of water either to switch to different irrigation activities or, assuming water is tradable between farmers, sell their water to willing buyers. USDA data provide us with a set of initial conditions for irrigation water use. Optimal allocations within USAGE-TERM will change as water scarcity worsens and the marginal product of water rises.

5.8.1 Reducing Surface Water Usage by 40 Percent and Pumping to Limit the Shortfall to 10 Percent

Groundwater accounts for about one-third of California's water requirements in a normal year (Chappelle et al. 2015). Howitt et al. (2015) estimated that groundwater usage rose by 6.2 million acre-feet and surface water usage fell by 8.7 million acre-feet in 2015 in California relative to a normal year, resulting in reduced water usage of 2.5 million acre-feet or about 10%. Groundwater in California is more important than in the Murray-Darling Basin. This marked difference in water sources has implications for both how a drought scenario is modeled and for a suitable policy plan.

In modeling drought, we also account for rainfall deficits. To illustrate, every inch of effective rainfall below average is equivalent to a water shortfall on every 12 acres of one acre-foot. Rainfall shocks in the scenario are based on estimates of rainfall deficits shown in Table 5.4 for July 2014 to June 2015. In addition to the modeled rainfall deficit, the shortfall of 2.5 million acre-feet is allocated among counties in proportion to annual crop water requirements of a normal year. The other

[8]Note that in USAGE-TERM, acre-feet are used in satellite accounts for water, not ML as in TERM-H2O, reflecting the respective units of USDA and ABS data. See http://quickstats.nass.usda.gov/.

Table 5.4 Effective rainfall deficit and irrigation cuts by county

	Average effective rain (inches)	Effective July 2014-June 2015 (inches)	Deficit (inches)	Irrigation cuts ('000 acre-feet)
ButteCA	25	10	15	41
ColusaCA	10	9	1	79
FresnoCA	7	4	3	263
GlennCA	13	10	3	54
MercedCA	6	2	4	142
KernCA	6	1	5	182
KingsCA	2	2	0	136
TulareCA	5	1	4	117
YoloCA	17	6	11	51
SanJoaquinCA	10	3	7	116
StanislausCA	7	3	4	94
MaderaCA	6	1	5	33
Rest of Calif.	13	5	8	1191

Source http://www.usclimate.com (accessed 28 August 2015)

shocks imposed in the scenario are to depict increased pumping costs arising from the extraction of an addition 6.2 million acre-feet of water. At $100 per acre-foot (Howitt et al. 2015), these additional costs total $620 million. In order to depict limited water trading possibilities, hay & forage and other agriculture water in each region are not transferable to other agricultural activities.

Table 5.5 shows the macroeconomic impacts. The macro impacts are relatively modest, with real GDP falling by 0.83% and real consumption, boosted by varying terms-of-trade gains across the 12 counties, lessening the losses in spending power. There are two costs associated with groundwater extractions that are not included in this scenario. First, additional groundwater extractions have required substantial investments in new or deeper wells. Only the marginal pumping costs are included in the scenario. We would expect present period investment to result in foregone consumption in the future, in order to service the debt incurred in financing the investment. But this version of USAGE-TERM was run in comparative static rather than dynamic mode. Consequently, the model does not account for dynamic linkages between investment and capital, and between international trade balances and net foreign liabilities.

Dynamics would also have been useful in the Californian context, to capture some of the longer term impacts of investment in boring of wells in response to the four year drought. But even a dynamic model would not have told the full story of the response. The other cost not modelled is the impact on sustainability: the more existing groundwater is pumped, the higher will be the costs of extracting more water as aquifer levels fall. Groundwater pumping is causing land subsidence, thereby

Table 5.5 Macroeconomic Impacts on 12 counties and terms-of-trade impacts, "Observed" scenario

12 counties	Macro results	Terms-of-trade by region		Water trading price $/ac-ft
Real consumption	−0.51	ButteCA	2.9	278
Real investment	−0.04	ColusaCA	0.7	107
Real GDP	−0.83	FresnoCA	0.3	337
Employment	−0.37	GlennCA	0.1	119
Real wage	0	MercedCA	2.1	571
GDP deflator	0.04	KernCA	0.4	259
CPI	−0.30	KingsCA	0.4	259
		TulareCA	0.8	361
		YoloCA	0.2	490
		SanJoaquinCA	0.5	452
		StanislausCA	−0.2	690
		MaderaCA	1.0	713

(% change from base case)

damaging property and the environment. Moreover, excessive pumping depleted the water supplies of some towns. This scenario does not depict the substantial costs of running out of town water.

5.8.2 *Why Water Pricing and Trading Would Help in California*

By 2015, there was some optimism in California that an El Nino would end the drought. But the policy problems remain. What benefits, for example, might arise from water metering and trading? If all water used in irrigation activity were metered, including groundwater, it would contribute to water being rationed according to scarcity. Metering and other reforms that enable water trading would encourage the movement of water from users with low average products of water to users with higher average products. In addition, metering of groundwater may reduce investments in perennials. In many irrigation regions around the world, there are physical constraints to trade, but in much of California's irrigation areas, the main constraints appear to be institutional. Metering would be most effective if the total groundwater extractions were limited to a volume that enables replenishment of aquifers in wetter years and drawing down in drier years, as proposed by Howitt and M'Marete (1991) a generation ago. A potential incentive that faces irrigators at present who invest in a new well is that instead of using it in dry years, they may perceive that they increase their return on the investment by using it in all years, to the detriment of sustainable management.

Water stress was not confined to farms, with water restrictions being imposed on urban users. The town of East Porterville, CA, for example suffered excessively from drought due to wells drying up. Prior to the prolonged drought, no authority recognized the vulnerability of the town's water supply. Various state and utility authorities planned to connect dry homes to mains water supply, and this finally happened in August 2016 (Stevens 2016).

5.9 Conclusion

Australia's Murray-Darling Basin and California's Central Valley contain two examples of irrigation schemes in developed countries. In each, severe droughts have exposed major issues arising from water allocations. In the Murray-Darling Basin, significant reforms prior to the millennium enabled water trading, which played an important role in helping the basin cope with drought. While water trading was a success, the objective of the Water Act 2007 to redress unsustainable water usage in the basin remains unrealized. Although the Commonwealth Government committed over $12 billion to the scheme, too much has been earmarked for water infrastructure upgrades that are yielding low benefits. Such upgrades are an inefficient means of providing regional economic benefits relative to alternative public expenditures. Purchasing water rights from farmers for environmental purposes is cheaper than aiming to increase environmental water through infrastructure upgrades.

Institutional arrangements in California lag behind those in Australia. Land and water ownership are not disentangled, so that water trading, if it occurs, proceeds in volumes far short of those that may occur if water were diverted to its most valuable uses during drought. Excessive drawing on groundwater has created a crisis in some communities. The externalities arising from the actions of irrigators in response to drought is a signal for public intervention. Analysts may ponder whether major institutional reforms will eventuate in California.

References

ABS (Australian Bureau of Statistics) (2008) Water and the Murray Darling Basin: a statistical profile, 2000–01 to 2005–06. Catalogue 4610.0.55.007

ABS (Australian Bureau of Statistics) (2009a) Water use on Australian farms, 2007–08. Catalogue 4618.0

ABS (Australian Bureau of Statistics) (2009b) Value of agricultural commodities produced, 2007–08. Catalogue 7503.0

Adamson D, Loch A (2014) The latest Murray-Darling plan could leave farmers high and dry. https://theconversation.com/the-latest-murray-darling-plan-could-leave-farmers-high-and-dry-27450. Accessed 31 Aug 2016

Australian Government (2007) A national plan for water security. http://nailsma.org.au/sites/default/files/publications/national_plan_water_security2007.pdf. Accessed 10 Sept 2016

Chappelle C, Hanak E, Mount J (2015) Reforming California's groundwater management. PPIC Water Policy Center. http://www.ppic.org/main/publication_show.asp?i=1106. Accessed 9 Sept 2016

COAG (1994) Water reform framework. http://webarchive.nla.gov.au/gov/20130904083606/. http://www.environment.gov.au/water/publications/action/policyframework.html. Accessed 9 Sept 2016

Davidson B (1969) Australia wet or dry? The physical and economic limits to the expansion of irrigation. Melbourne University Press, Melbourne

Delaney A (2010) Massive water meeting for Griffith. http://www.abc.net.au/local/stories/2010/10/15/3039023.htm. Accessed 31 Aug 2016

Dixon P, Rimmer M (2010) You can't have a CGE recession without excess capacity. Econ Model 28:602–613

Dixon P, Parmenter B, Sutton J, Vincent D (1982) ORANI: a multisectoral model of the Australian economy. Contributions to economic analysis 142, Amsterdam, North-Holland

Dixon P, Rimmer M, Wittwer G (2011) Saving the Southern Murray-Darling Basin: the economic effects of a buyback of irrigation water. Econ Rec 87:153–168

Howitt R, M'Marete M (1991) Well set aside proposal: a scenario for ground water banking. Calif Agric 45(3):6–8

Howitt R, MacEwan D, Medellín-Azuara J, Lund J, Sumner, D (2015) Economic analysis of the 2015 drought. UC Davis Center for Watershed Sciences ERA Economics. https://watershed.ucdavis.edu/files/biblio/Final_Drought%20Report_08182015_Full_Report_WithAppendices.pdf. Accessed 26 Sept 2018

Kagi J (2016) Ord River project plagued by cost blowouts, delays, poor planning, auditor-general finds. http://www.abc.net.au/news/2016-09-07/ord-river-project-plagued-by-cost-blowouts-delays-report/7823422. Accessed 26 Sept 2018

NSW Irrigators' Council (2010) Submission to productivity commission: market mechanisms for recovering water in the Murray-Darling Basin. http://www.pc.gov.au/inquiries/completed/murray-darling-water-recovery/submissions/sub032.pdf. Accessed 26 Sept 2018

Shields M (2016) Riverina mayors call for water policy rethink after SunRice slashes jobs. http://www.abc.net.au/news/2016-01-29/riverina-mayors-call-for-water-policy-rethink/7123112. Accessed 26 Sept 2018

Stevens M (2016) After years without water, taps are turned on in East Porterville. http://www.latimes.com/local/lanow/la-me-ln-east-porterville-20160819-snap-story.html. Accessed 26 Sept 2018

Wheeler S, Zuo A, Bjornlund H (2014) Investigating the delayed on-farm consequences of selling water entitlements in the Murray-Darling Basin. Agric Water Manage 145:72–82

Wittwer G (2011) Confusing policy and catastrophe: Buybacks and drought in the Murray–Darling Basin. Ec Papers 30:289–295

Wittwer G (2015) From almond shaming to water trading: CGE insights into managing California's drought. Centre of Policy Studies working paper G-258

Wittwer G (ed) (2017) Multi-regional dynamic general equilibrium modeling of the U.S. economy. Springer, Dordrecht

Wittwer G, Banerjee O (2015) Investing in irrigation development in North West Queensland, Australia. Aust J Agric Res Econ 59(2):189–207

Wittwer G, Dixon J (2013) Effective use of public funding in the Murray-Darling Basin: a comparison of buybacks and infrastructure upgrades. Aust J Agric Res Econ 57:399–421

Wittwer G, Griffith M (2011) Modelling drought and recovery in the southern Murray-Darling basin. Aust J Agric Resour Econ 49:342–359

Wittwer G, Griffith M (2012) The economic consequences of a prolonged drought in the Southern Murray-Darling Basin. In: Wittwer G (ed) Economic modeling of water. Springer, Dordrecht, pp 119–141

Prof. Glyn Wittwer is a regional dynamic CGE modeling expert. He has played a major role with Mark Horridge in developing databases for TERM versions in several countries. He edited the Springer volumes *Economic Modeling of Water* (2012) and *Multi-regional Dynamic General Equilibrium Modeling of the U.S. Economy* (2017), and contributed the majority of chapters in both volumes. Glyn is a GTAP Research Fellow from 2017 to 2020. He has extensive consulting experience. His list of projects includes dynamic, multi-regional CGE modeling in Australia, China and the United States. These includes modeling of the impacts of major dam and transport projects, drought and water trading, flood, hypothetical plant disease scenarios, productivity scenarios, wine tax scenarios, major mine construction projects, industry closures and the aftermath of civil disruption.

Chapter 6
Any Rain on Victoria Lake Is Only a Drop in the Bucket: A CGE Analysis of the Effects of Water Shortages on Food Security in Uganda

Jan H. van Heerden, Elizabeth L. Roos, Philip D. Adams and N. Kilimani

Abstract Using a dynamic Computable General Equilibrium (CGE) model of Uganda, we simulate the effects of water shortages and their impact on agricultural production and the broader Ugandan economy. It is expected that Ugandan crop production will be hit hard over the next forty years by increasing temperatures and decreasing precipitation associated with climate change. We use forecasts from the literature for ten specific crop outputs to simulate the effects of weather-related agricultural disruption on the rest of the economy. We are particularly interested in the effects on food security in Uganda. We find that Uganda is far from being food secure, and hypothesize that if the pessimistic forecasts from the literature were to become true, then the situation would deteriorate significantly. The adaptation strategies for Uganda should focus on the following: the country should diversify agricultural production to include more hardy crops; build transport infrastructure to improve access to the available food to all citizens, but also to enable the expansion of the manufacturing industry, which is dependent on trade between regions; improve sanitation conditions significantly in both rural and urban areas, and develop water

The work described in this chapter was carried out at the Centre of Policy Studies and the University of Pretoria. The original UgAGE model was developed for the Ugandan Ministry of Finance, Planning and Economic Development. We kindly acknowledge their support. The model used in this chapter has been adapted to focus on agriculture and food security. The views expressed are those of the authors and should not be attributed to the Ministry of Finance, Planning and Economic Development.

J. H. van Heerden (✉)
Department of Economics, University of Pretoria, Pretoria, South Africa
e-mail: jan.vanheerden@up.ac.za

E. L. Roos · P. D. Adams
Centre of Policy Studies, Victoria University, Melbourne, Australia
e-mail: Louise.Roos@vu.edu.au

P. D. Adams
e-mail: philip.adams@vu.edu.au

N. Kilimani
Department of Economics, Makerere University, Kampala, Uganda
e-mail: nkilimani@cis.mak.ac.ug

G. Wittwer (ed.), *Economy-Wide Modeling of Water at Regional and Global Scales*, Advances in Applied General Equilibrium Modeling,
https://doi.org/10.1007/978-981-13-6101-2_6

infrastructure so that households and agricultural industries can gain better access to an increasingly scarce resource.

Keywords Climate change · Food stability · Crop production · Adaptation strategies · Uganda

6.1 Introduction

Lake Victoria is the second largest fresh water lake in the world. It spans 68,800 km^2, and has a shoreline of 7000 km. It contains the seemingly infinitely large source of the mighty Nile River. The Nile flows through ten upstream countries before reaching Egypt and before it runs into the Mediterranean Sea.

Uganda is a relatively small fertile country on the equator in Africa and it is fortunate enough that almost half of Lake Victoria lies within its borders. About 20 per cent of the Ugandan border runs through the middle of the lake.

Uganda also has other lakes on or within its borders. From Lake Victoria, the Nile runs through the countryside into Lake Albert, which forms a large portion of the western border with the DRC. From there the river crosses the border into Southern Sudan. A third large lake is Lake Kyoga, which is quite extensive towards the upper middle part of the country, while Lake Edward lies in the south-western corner of Uganda in the Queen Victoria National Park.

If one has visited the south-eastern parts of Uganda around Kampala or Entebbe, and experienced some of their torrential rainstorms and the resulting ungovernable muddy roads, you might make the mistake that Uganda does not and probably never will have water supply problems.

Even so, Uganda is also the place where the UN World Food Program (WFP) provided emergency food aid for more than 40 years until 2012, when it scaled back its operation to provide more targeted, voucher, and cash-based assistance. Food shortages remain a problem to this day because of drought-affected harvests.

In this chapter, we examine the state of food security in Uganda and explore the effects of possible rainfall decline over the coming decades through to 2050. We conclude with possible adaptation strategies, without modelling them specifically.

Section 6.2 examines the current state of food security in Uganda. The effects of future water shortages due to climate change on specific crops are examined in Sect. 6.3. The Computable General Equilibrium (CGE) model used for the analysis is described in Sect. 6.4; Sect. 6.5 covers simulation design, assumptions and results, followed by conclusions in Sect. 6.6.

6.2 Food Security in Uganda

6.2.1 Defining Food Security

The Food and Agricultural Organization of the United Nations (FAO) claims that by 1980 there had been more than 200 definitions of food security published in research and policy usage (FAO 2003). The initial focus of attention was primarily on food supply problems in terms of (i) the availability of food and (ii) the price stability of basic foodstuffs nationally and internationally. The 1974 definition of food security was given as "the availability at all times of adequate food supplies of basic foodstuffs" (p.1).

The world food crisis of the mid-1970s led to a redefinition of food security to stress the importance of potentially vulnerable and affected people as a critical aspect.[1] Even if there was a sufficient supply of food, there could be a lack of effective demand due to poverty and low incomes. The definition of food security was therefore expanded to include aspects of both supply and demand of food. Moreover, the "food" in "food security" was re-defined to include a certain level of nutritional ingredients.[2]

> Food security [is] a situation that exists when all people, at all times, have physical, social and economic access to sufficient, safe and nutritious food that meets their dietary needs and food preferences for an active and healthy life (FAO 2002).

This definition is clearly more expansive than the one stressing mere *availability* of basic foodstuffs. The physical and economic *access* points towards the demand side of consumers who should have the means to buy food – food of appropriate *quality* that is much more than basic foodstuffs: it should be sufficient, safe, nutritious, and should meet very specific dietary preferences. The goal of food security is therefore not only to survive, but also to be able to lead an active and healthy life. Finally, there should be *stability* of food security so that households or individuals could utilize the food at all times.

We now discuss the four aspects of food security with respect to Uganda.

6.2.2 Food Availability

The literature indicates that Ugandans frequently do not have enough food to eat. A recent report by the National Planning Authority indicates that many households in Uganda have been suffering from food insecurity, due to a lack of income to buy food, or due to droughts that made it difficult to produce their own food:

---

[1] In the early 1970s there was scarcity in the world grain market, soaring prices and famines spread over several countries of Asia and Africa. A world food conference was held in 1974 to address world hunger (FAO 2003).

[2] See https://www.wfp.org/node/359289 accessed 24 September 2018.

> On a national level, throughout the period 2009/10–2015/2016, 16 per cent of the households were chronically undernourished and only 4 per cent of the households remained food secure. Of great concern is the rather high vulnerability to food insecurity, with 80 per cent of the households having suffered transient food insecurity. At the regional level, the northern region was home to nearly 40 per cent of the chronically food insecure households, followed by the eastern region at 27 per cent.

The report further notes that:

> For some households, food shocks were recurrent but varied in severity for the different periods spanning the survey. The households reporting a high prevalence of food shocks were in rural areas and were more likely to be reliant on agricultural sources. Such households have lower average wealth levels to act as buffers in difficult times.
>
> The reasons behind food shocks were primarily related to food production inadequacies due to climatic shocks. Drought/poor rains were the most cited cause, particularly for rural households. This cause is followed by income insecurity, particularly in urban areas, and high food prices, which equally affect both rural and urban households. A large portion of households also reported they did not plant enough to ensure availability of food stocks (NPA 2017, p. 59).

6.2.3 Utilization of Food

Food security was defined above as "access to safe and nutritious food", amongst others.

Water.org claims that more than 40% of people living in the greater Kampala area live in unplanned settlements and that only 17% of them have access to piped water.[3] If this is true of the greater Kampala area, which is situated adjacent to the Victoria Lake, then it must be worse in the rural areas. Many people therefore use springs and other surface sources that are highly contaminated for this reason, and because the price of safe water is very high (three time higher than in planned urban communities). The problem is also getting worse in Kampala because of migration by the high growing population to the informal settlements. The authors claim that "more than half of all residents in Uganda do not have access to improved sanitation, sharing overcrowded pit latrines or practicing open defecation," which cause frequent outbreaks of waterborne diseases such as cholera and dysentery.

The World Bank confirms the desperate state of affairs in Uganda and claims that 13.8 million people use unsanitary or shared latrines, while 3.2 million have no latrine at all and defecate in the open.[4] (World Bank 2012). In some districts of Uganda, the situation is very bad: according to Wilfred Okello, a water and sanitation specialist, in the Karamoja sub-region, out of 128 villages in Napak District, only four have latrines. In Kaabong, out of 287 villages only six have latrines. For the case of Kotido,

---

[3]See http://water.org/country/uganda/ accessed 24 September 2018.

[4]See https://www.zaragoza.es/contenidos/medioambiente/onu/825-eng-v14.pdf accessed 24 September 2018.

there are 187 villages and of these, only two have pit-latrines. Nakapiripiriti, on the other hand, has only one village that has latrines (Nakibuuka 2016).

According to the WHO, "most diarrheal deaths in the world (88 per cent) are caused by unsafe water, sanitation, or hygiene. More than 99% of these deaths are in developing countries, and about 8 in every 10 deaths are children"[5] (WSP and Unicef 2014). Diarrhoea obliges households to spend money on medicine, transportation, health facility fees, and more and can mean lost work and wages among working household members.

The Water Institute at the University of North Carolina in Chapel Hill visited Uganda for a case study of community-led total sanitation (CLTS) implementation, focusing on the roles and responsibilities of local actors. They are happy that the Ugandan government has policies in place as well as a dedicated budget to support these goals, but report that many respondents cited termites and heavy rains as reasons why toilets of poor quality construction would collapse. The Institute found that there is a large shortage of skilled masons that are able to build durable toilets and decided to identify and train 78 masons in one district.[6]

We claim that it is highly unlikely to fulfil the requirement of having access to "safe and nutritious food" under the circumstances described above, for obvious reasons. If people do not or cannot wash their hands, they will probably not be able to work with food in a hygienic way. If they have access to food, they are likely to contaminate the food.

6.2.4 *Food Stability*

USGS has prepared a brief report in which they examine Ugandan rainfall and temperature trends for the last 110 years (1900–2009) using observations from 57 rain gauges and 5 air temperature stations for the two primary rainy periods of Uganda, namely March-June and June-September (Fig. 6.1). The report summarizes a multi-year effort by the USAID Famine Early Warning Systems Network (FEWS NET). It concludes that (i) "Both spring and summer rains have decreased in Uganda during the past 25 years" and (ii) "The magnitude of observed warming, especially since the early 1980s, is large and unprecedented within the past 110 years, representing a large (2 + standard deviations) change from the climatic norm" (p.1).[7]

Figure 6.1 shows how rainfall and temperature have changed during the past 50 years in Uganda. The dark brown area in the bottom left map is receiving 100mm less rain than before, while the reddish area in the north-east of the bottom right map is experiencing 1.5 degrees C increase in temperature, etc.

---

[5] See https://www.wsp.org/sites/wsp.org/files/publications/WSP-Uganda-CFD-Profile.pdf accessed 24 September 2018.

[6] See https://waterinstitute.unc.edu/files/2015/12/learning-series-uganda-learning-brief-2015-12.pdf accessed 24 September 2018.

[7] See https://pubs.usgs.gov/fs/2012/3062/FS2012-3062.pdf accessed 24 September 2018.

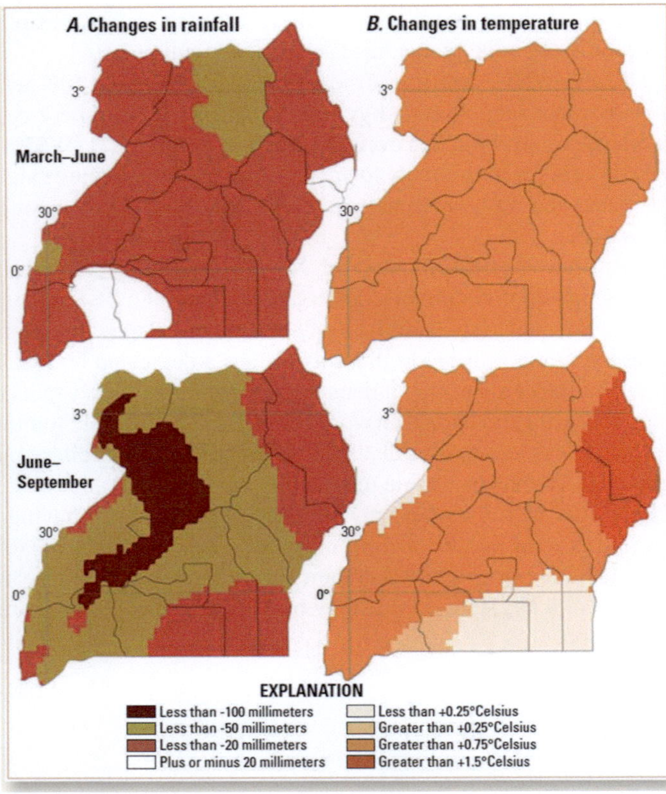

Fig. 6.1 Observed (1960–2009) changes in March-June and June-September rainfall and temperature in Uganda. *Source* USGS (see footnote 8)

Figure 6.2 tells the story very clearly in times series form. The 2000–2009 rainfall has been, on average, about 8 per cent lower than rainfall between 1920 and 1969, with the June-September rainfall declining for more than two decades now, while the March-June decline starting a bit later.

The temperature in Uganda increased by more than 0.8 degrees Celsius (°C) on average, and up to 1.5 °C in the north east of the country.

USGS concludes that "a decrease in locally produced food because of reduced crop harvests and pastures will have a significant impact on food security". In particular, warming temperatures may adversely affect coffee production, which is an important cash commodity in Uganda: "…coffee plants fare poorly in areas with average air temperatures greater than 24 °C". Moreover, USGS is alarmed at the high population

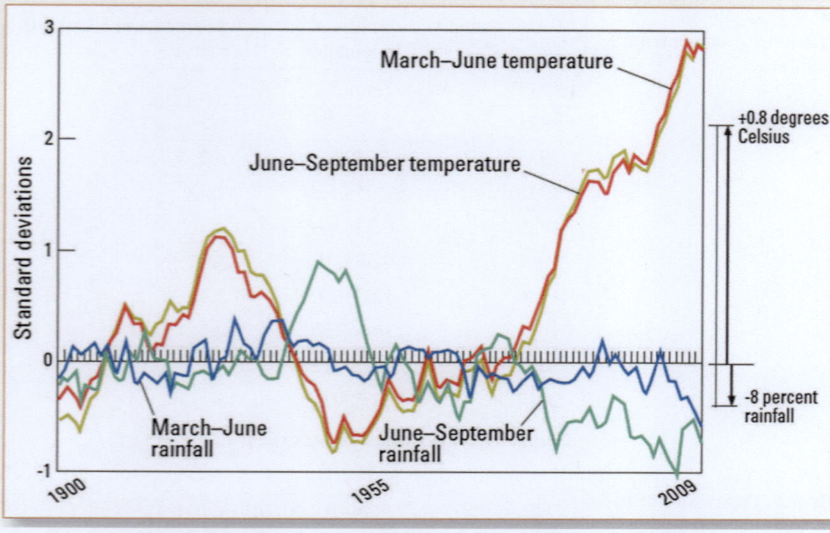

Fig. 6.2 Smoothed 1900–2009 March–June and June–September rainfall and air temperature time series for crop-growing regions. *Source* USGS (see footnote 8)

growth rate of 3.2 per cent of Uganda[8] (CIA 2016), which might lead to a "per capita decline in cereal production of 35 per cent by 2025".

Food stability means that people must have access to adequate food at all times and not risk losing access to food as a consequence of sudden shocks (e.g. an economic or climatic crisis) or cyclical events (e.g. seasonal food insecurity). The trends described here convince us that Uganda does not have food stability now and that the situation may worsen in the near future.

Statistics provided by UBOS on (i) population growth in Uganda and (ii) the prevalence of food shocks among the people of Uganda confirm the notion that food provision would become more unstable in the near future.

Table 6.1 shows the population size and growth rates over time, while Table 6.2 shows the percentage of all households in all regions that have been experiencing food shocks. The Ugandan population is growing at an alarming rate and it does not seem as if a situation of a stable food supply to all households is going to be achieved soon, especially in the northern areas.

[8]See https://www.cia.gov/library/publications/the-world-factbook/geos/ug.html accessed 24 September 2018.

Table 6.1 Ugandan population size and annual growth rates

Census year	Total population	Average annual growth rate
1911	2,466,325	
1921	2,854,608	1.5
1931	3,542,281	2.2
1948	4,958,520	2
1959	6,536,616	2.5
1969	9,535,051	3.9
1980	12,636,179	2.7
1991	16,671,705	2.5
2002	24,227,297	3.2
2014	34,634,650	3

Source Uganda Bureau of Statistics

Table 6.2 Prevalence of food shock experiences among households (%)

	2009/10	2010/11	2011/12	2013/14	2015/16
National	54.9	38.6	32.3	39.8	26.3
Urban-Rural					
Rural	6.1	41.8	32.7	45.2	29.2
Urban	36.5	22.4	30.7	24.2	17.7
Region					
Central	51.8	39.2	32.4	35.9	23.5
Eastern	49.1	39.3	34.8	37.7	19.9
Northern	73	50.7	26.8	46.6	51.8
Western	51	25.1	34.5	41.8	17.5

6.3 Climate Change in East Africa and Its Likely Effects on Certain Agricultural Crops in Uganda

According to Hepworth (2010), "the Ugandan economy and the wellbeing of its population are particularly vulnerable to climate change and this is evidenced by the widespread damage and hardship imposed by regular drought and extreme rainfall events under current climate variability" (p. 13).

In the modelling reported in Sect. 6.4, we use a CGE model of Uganda to simulate a climate-change (pessimistic) scenario for the main crops of Uganda. The main inputs to the simulation are predictions from the literature on possible agricultural output changes by 2050 in different crops.

6.3.1 Bananas

According to the FAO, 'banana is the most important food crop in Uganda and Rwanda, where the annual per capita consumption exceeds 135 kg. per annum (FAO 2013a). This amounts to 3–11 bananas per person per day. In Uganda, the local word for bananas–matooke– means 'food'.

According to Adhikari et al. (2015) "quantitative measures of banana yield under future climate change are limited". The literature also differs markedly on the forecasts for future yield. Biodiversity International researched the possible effects of climate change on world banana production. Calberto, who led the study, said that by 2070 the land suitable for banana production would increase by 50%, due to warmer climates. However, according to co-author, Staver, 'higher temperatures mean an increase in water demand of 12–15%.' (Calberto et al. 2015, p. 265). Adhikari et al. quotes Van Asten et al. (2011) who claimed that "in the East-African highlands, drought-induced yield reduction on rain fed bananas can reach up to 65% compared to wetter areas (2015).

Similar to our Cassava forecast (Sect. 6.3.3), we only simulate the pessimistic outlook for Bananas in our modelling simulations.

6.3.2 Beans

According to Wortmann et al. (1998) beans are the second most important source of protein after maize in eastern and southern Africa, and the third most important source of calories after cassava and maize. Concerning the effects of climate change on soybean production, Wortmann et al. (1998) contend that rainfall variability and soil moisture, rather than rising temperature, are the crucial determining factors.

We use four sources from the literature to determine the shock to be applied to the productivity of bean production in the climate change simulation run.

Lobell et al. (2008) used 20 General Circulation Models (GCMs)[9] and projected a slight decrease of less than 5 per cent in bean production in East Africa by 2030 as compared to 1998–2002. At this rate of growth, the decrease would amount to about 10 per cent by 2050.

Thornton et al. (2009) used the HadCM3 and ECHam4 models and the A1F1 and B1 storylines and forecast decreases of up to 20% in production of beans by 2050, in 56% of the land under bean production in East Africa. In some highland areas production could increase, but in arid and semi-arid areas production would decrease. If 56% of the crops decrease by 20%, the total decrease would amount to 11.2%. With production increasing in some areas, the overall decrease would therefore be less than 11.2%.

---

[9]Numerical models representing physical processes in the atmosphere, ocean, cryosphere and land surface, used to simulate the response of the global climate system to increasing greenhouse gas concentrations.

Tatsumi et al. (2011) used the IGAEZ[10] method under the A1B storyline to estimate a decrease in bean production of 12% by 2090, while Boote et al. (1998) found a decrease of 3–10% across all the IPCC storyline scenarios towards the end of the century.

Since the estimates from Thornton, Tatsumi and Boote are similar to that of Lobell et al. we will use a production shocks of between −3 and −10% for bean production.

6.3.3 Cassava

Cassava is considered to be the most important product in terms of caloric intake and the most important in terms of total production in East Africa (Fermont 2009). According to Jarvis et al. (2012) "the crop is more resilient to climate change due to its tolerance of high temperatures and intraseasonal drought". In a series of model simulations, these authors concluded that cassava would perform as least as well as maize, sorghum, millet, beans, potatoes or bananas in the presence of climate changing conditions, "making it a potential candidate to ensure future food security in the region."

The long run forecasts of Cassava production differ vastly from each other. Lobell et al. (2008) projected a decrease of 8% in output by 2030, which would be synonymous to a 15.8% decrease by 2050. Tatsumi et al. (2011) forecast a 10% increase in Cassava yield by 2090, while IFPRI forecast increases between 40 and 100% in Uganda, Rwanda and Malawi, using the IMPACT model.

Even though the literature differs from large decreases to large increases in Cassava production because of climate change, we have decided to model only the pessimistic forecast for Cassava and other yields. For the pessimistic forecast we will use Lobell et al. (2008), which differs from the optimistic forecast of Tatsumi et al. (2011).

6.3.4 Coffee and Tea

Arabica coffee grows at altitudes above 1000 m and prefers cool temperatures. Global warming will force coffee growers to move to higher altitudes that they are now working at, if land is available, so, according to Moat et al. (2017), by the end of this century "the current coffee-growing niche of Ethiopia could decrease by 39–59% without interventions". This amounts to a decrease of between 14 and 20% by 2050.

Bunn et al. (2015) predict that Robusta coffee will experience a loss of 16–30% in Eastern Africa by 2050, while Arabica coffee production will decrease between 10 and 30% (p. 97). Their predictions for the rest of the world are much worse for

[10]Improved Global Agro- Ecological Zones method.

both varieties. We will use a weighted average[11] of the Bunn et al. predictions for our pessimistic forecast.

The outlook for tea in Uganda looks much worse than for coffee. The International Center for Tropical Agriculture predicts that more than half of the areas used for tea production in Uganda will become unsuitable due to climate change. They do not predict the exact decrease in production, and since we are grouping coffee and tea together in our model, we implicitly make the strict assumption that tea production will decrease by as much as coffee production.

6.3.5 Cotton

Although Adhikari et al. (2015) expect the productivity of cotton production to decline with climate change, they do not provide any specific predictions for it. They bemoan the fact that the quantitative analysis of future cotton production is sparse.

The ITC (2011) did perform a world-wide study of the impact of climate change on cotton production, but is equally hesitant to provide specific numbers for predicted changes in output. Insightful conclusions coming from the ITC report are that they predict increases in the production of cotton for both China (p. 14) and the United States (p. 20), due to increasing temperatures and resulting longer crop seasons.

In light of these optimistic forecasts we assume that cotton production in Uganda will not necessarily be affected negatively by climate change. We model this prediction by leaving cotton production endogenous in the model, with the implication that we do not forecast it specifically, but allow the industry to react to our model simulations.

6.3.6 Maize

Broughton et al. (2003) consider maize to be the most important source of dietary protein and the second most important source of calories in Eastern and Southern Africa. Maize is also one of the crops for which the possible climate change impacts have been widely studied (Adhikari et al. 2015, p. 115). They conclude, "Overall, maize production will decrease under future climate scenarios, though the degree of impact differs among simulations" (p. 116).

Lobell et al. (2011) found that "each degree day spent above 30 °C has been found to reduce final maize yield by 1% even under optimal rain fed conditions." Lobell and Field (2007) found that a 1 °C increase above the norm would reduce the maize yield by 8.3%, while Brown (2009) contended that the number would be −10%.

---

[11] Uganda coffee production consists of 57.2% Arabica and 42.8% Robusta, which we use as weights here.

6.3.7 Millet

Adhikari et al. (2015) conclude that millet is more resilient to climate change than maize or wheat but less resilient than sorghum (p. 119). We therefore use numbers between our respective pessimistic and optimistic forecasts for maize on the one hand and sorghum on the other. Millet resorts under "Other cereals" in the model and we therefore assume that the remaining small cereals would react in the same way as millet.

6.3.8 Rice

According to Adhikari et al. (2015) the projections on future rice productivity differ among studies, but most studies predict small or insignificant changes in rice production. Lobell et al. (2008) predict a small increase by 2050 (less than 5%), Ringler et al. (2010) predict a mere 0.24% increase for East Africa, and Tatsumi et al. (2011) predict only a small gain. In the light of these predictions we will assume that rice productivity will stay on the baseline.

6.3.9 Sweet Potato and Potato

Uganda is the third largest producer of sweet potato in the world (Adhikari et al. 2015, p. 121). Unfortunately, the information on the future of the crop in the presence of climate change is sparse. Ringler et al. (2010) predict an increase of 1.06% by 2050, while Tatsumi et al. (2011) predict a decrease of 7% by then. Adhikari considers the studies on the topic "inconclusive".

As with sweet potato there are very few studies on the impacts of climate change on potato production in Eastern Africa. Moreover, the studies differ vastly in their predictions: IFPRI forecast output to double or even triple by 2050, while Jarvis et al. predicted a decrease of 34% by 2050, and Tatsumi et al. a decrease of 8%.

Since we have grouped all root crops together in our model, we will simulate a pessimistic outlook of −8%.

6.3.10 Sorghum

Taylor (2003 cited in Adhikari et al. 2015, p. 119) claims that sorghum has an inherent ability to resist drought and resist periods of high temperatures. Adhikari et al. conclude that "sorghum is more resilient to climate change than maize or wheat and will be minimally impacted (< 5%) by the middle of this century (p. 120)."

We therefore will use a decrease of 5% in production as our pessimistic scenario.

6.3.11 Wheat

Wheat is a cool season crop and "increasing temperatures shortens its growth period by accelerating phonological development, resulting in reduced yield" (You et al. 2005; Asseng et al. 2011 cited in Adhikari et al. 2015, p. 117). There is consensus in the literature that wheat has a lower optimal temperature than rice, maize, millet, cassava and sorghum (Liu et al. 2008) so that most simulation studies predict a greater impact on wheat crop yields compared to other crops in Eastern Africa. (Liu et al. 2008; Fisher 2009; Nelson et al. 2009; Ringler et al. 2010).

The forecasts of yield changes look quite bleak for wheat: Fisher (2009) using HadCM3 along with the A2 storyline forecasts a 63% decrease in output by 2050. CSIRO predicts a drop of 48% by 2050, while Nelson et al. (2009) only forecast a decrease of 34–36%. Tatsumi et al.'s forecast is much smaller. We will use Fisher's forecast as our pessimistic view of wheat forecast.

6.3.12 Livestock

Thornton et al. (2009) claim that "the intersection of climate change and livestock in developing countries is a relatively neglected research area," and Thornton has been involved in many research articles on this topic. In one of these papers Galvin et al. (2004) studied a Maasai pastoralist community in the Ngorongoro and conclude, "rising human population along with many land tenure and land-use changes have squeezed pastoral livestock onto land areas that are too small to be sustainable for pastoral production." Warmer temperatures bring more disease epidemics, while livestock starvation is associated with recurring droughts. Galvin et al. come to the conclusion that we experience a rising human population in east Africa who are dependent on a "stable or declining livestock population" (p. 2). Since there is no consensus in the literature about the future trends in livestock production, we do not directly forecast its future growth. We let the model calculate how the livestock industry would react to the shocks we apply to other agricultural crops.

6.3.13 Fisheries

Allison et al. (2009) studied 132 countries and found Uganda to be among the top 5% most vulnerable of all countries regarding their fisheries, but unfortunately also among the countries with the lowest adaptability potential.

Table 6.3 Assumptions for agricultural production in the rainfall deficient climate change scenario (% changes relative to no climate change)

Crop	Pessimistic view (simulated)	% Shock per annum	Optimistic view (not simulated)	% Shock per annum
Annual crops	No shock	–	No shock	–
Coffee and Tea	−30	−0.658	−12.6	−0.297
Wheat	−63	−1.229	−34	−0.734
Maize	−10	−0.239	−8.3	−0.2
Sorghum	−5	−0.122	0	0
Other cereals	−8	−0.193	−4	−0.098
Pulses	−10	−0.239	−3	−0.074
Root crops	−8	−0.193	+8	0.193
Cassava	−16	−0.372	+10	0.239
Bananas (matooke)	−5	−0.122	+10	0.239
Livestock	No shock	–	No shock	–
Fishing	−10	- 0.239	0	0

6.3.14 Overall

Table 6.3 summarises the predictions of changes in crop production due to water shortage and temperature change associated with climate change that are used in the modelling. The changes are expressed as permanent percentage deviations away from a no climate change control. The table shows two sets of changes—pessimistic and optimistic. Only the pessimistic changes are modelled, as our interest lies in the effects of water shortages.

6.4 The Model and Database

6.4.1 Overview of Model Structure

We use a dynamic version of the UgAGE CGE model built by the Centre of Policy Studies. UgAGE is based on a framework adopted for models of many countries, as outlined in Dixon and Rimmer (2002a) and Dixon et al. (1982).

UgAGE models the production of 95 commodities by 95 industries. Three primary factors are employed by the industries, namely labour, capital and natural resources. Capital is assumed to be sector-specific, while occupation-specific labour is perfectly mobile across industries.

The model identifies 32 household groups defined by income level. Each household is modelled as a budget constrained maximizer of utility.

The model recognizes consumption of commodities by government and the details of various taxation instruments. As the default, it is assumed that all sectors are competitive and all markets clear.[12]

The second part of the model's theoretic structure relates to dynamics and includes two types of mechanisms:

- stock/flow accounting for the accumulation of industry-specific capital stocks via investment and depreciation; and
- lagged adjustment mechanisms in the labour market.

Capital accumulation is industry-specific, and linked to industry-specific net investment. Changes in industry-specific investment is linked to changes in industry-specific rates of return.

UgAGE includes a labour market mechanism that allows the labour market to move from a typical short-run scenario (employment adjusts with real wage held at its baseline level) to a long-fun scenario (real wage adjusts with employment held at its baseline level). With this mechanism operational in the policy simulation, short-run labour market pressures mostly manifest as changes in employment. In the long run, employment returns to baseline, with labour market pressures reflected in changes in real wages.

The third part of the model's theoretical structure describes transfers (flows) between various entities, not described in the first two parts. These transfers are included in the SAM. To facilitate use of SAM data, UgAGE includes equations describing transfers to and from:

- 32 household groups;
- 2 enterprise types (non-financial and financial enterprises);
- the government; and
- the rest of the world (ROW).

The intra-entity transfer data is not dependent on specific behaviour. Instead, the changes in the transfers between entities are linked to variables in the core model; for example, the transfers from the ROW to households follow nominal GDP.

[12]Other options are available, provided that there are data to support the non-default settings. For example, if there are data on economies or diseconomies of scale in certain industries, then this can be incorporated into the database and into the theoretical structure of the model. Similarly, if there are data suggesting imperfectly competitive market conditions, then this can also be incorporated into both the data and equation structure of the model.

6.4.2 UgAGE Database

The model's database includes information on 95 commodities and industries (of which 13 are agricultural industries), sourced either domestically or from abroad, and sold on the local or foreign market, or transferred between entities.

The database is calibrated to the 2010 Supply Use Table (SUT) of Uganda. Additional information, not captured in the SUT, is taken from the 2010 Social Accounting Matrix (SAM). The SUT and SAM data are not in the format required by the model.[13] Transforming published data into the required IO format, is a laborious task that is not reported upon here.[14]

The following interesting aspects of the Ugandan economy are observed in the database: On the income side of GDP, the cost of capital is more than double the cost of labour. The combined shares of natural resources (agricultural land and mining reserves) and capital account for 66 per cent of all factor inputs. This means that in the years immediately following a shock to the economy, with capital and natural resources largely fixed, the economy's generalized supply schedule is inelastic.

On the expenditure side of GDP, the largest component is private consumption, with a GDP-share of 72 per cent. Imports exceed exports, implying a deficit on the trade account of 3561 billion Shillings, nearly 9 per cent of GDP. The remaining components as a share of GDP include exports (17.8%), imports (−26.6%), investment (26.5 per cent) and government consumption (9.5%).

The Ugandan economy is concentrated around services and agricultural sectors. The services sector contributes 46.6 per cent of total value added while the agricultural sector contributes 28.6 per cent. The manufacturing industries only contribute 8.9 per cent to value added followed by construction (6.2%), utilities (2.9%) and mining (1.2%).[15] In terms of agriculture, pulses (3.9%), livestock (5.2%), cassava (2.7%), maize (2.3%), root crops (1.9%) bananas (2.3%) and forestry (4.2%) contribute the most to value added.

Industry-specific results are influenced by changes in demand and supply. The cost composition of an industry assists in explaining the change in supply while the sales structure and macroeconomic mechanisms are important in explaining changes in demand.

In the first year following a shock, capital is largely fixed in place by investments undertaken prior to the shock. It is assumed that the real wage rate is initially

[13]The model requires a core database with separate matrices for basic, tax and margin flows for both domestic and imported sources of commodities sold to domestic and foreign users. These separate matrices are not explicitly captured in the SUT. Instead they appear in an aggregate form e.g. the USE table shows the use of commodity c by user u valued at purchasers price. These values should be disaggregated into three separate matrices namely, the use of commodities valued at basic price, tax and margin flows. They then have to be disaggregated to show where these commodities are sourced from, either from the local market or imported from the ROW.

[14]See Roos et al. (2015) for a description of how such an IO database can be created from a SUT.

[15]Downstream manufactures such as textiles, meat processing, grain and bakery industries contribute little to GDP. Most agricultural commodities, including sorghum, cereals, pulses, rootcrops, cassava and bananas are sold mainly to private households.

sluggish. Hence, immediately following a shock, any increase (decrease) in output will typically be accommodated by an increase (decrease) in employment, holding technologies unchanged. Capital-intensive industries, therefore, have little scope to adjust their output in the short-run.

Over time, however, industries respond to a shock via capital adjustment. It is assumed that the real wage rate is flexible in the long-run, and moves to eliminate any short-run change in employment. Therefore, in the long-run, capital-intensive industries have greater scope to adjust their output relative to labour-intensive industries.

On the demand side, the initial sales structure as well as assumptions about macroeconomic variables impact on industry production in both the short and long-term. For example, if private consumption is held fixed, then there is little scope for industries selling their output to households to change their production. On the other hand, because export demand is assumed to be relatively elastic, there is much more scope for industries that are export-focused to change their production.

The costs shares in the database show that agricultural sectors use land as an input. Land is fixed by assumption and can therefore not be substituted for other inputs given changes in relative price. The share of capital in agricultural production cost is fairly high with the average capital share for the entire agricultural sector above 40 per cent. The share of locally and imported intermediate inputs is lower for the agriculture than for manufacturing. For example, Food, beverages and tobacco uses 57.9 per cent as locally sourced intermediate commodities during the production process, while the meat processing industry uses 93.3 per cent as locally sourced intermediate commodities during the production process.

The sales structure for commodities is also very informative. We note that commodities such as machinery and construction are used to create capital while public administration, education and health services are used by the government. Domestically produced rice, sorghum, other cereals, pulses, root crops, cassava and bananas are mostly consumed by households. Therefore, any negative shocks to these commodities, such as a prolonged drought, will impact households directly. Commodities such as wheat, maize and livestock are used as intermediate inputs in the grains industry and meat processing industry. Interestingly, we see that coffee beans are mainly exported while wheat sold on the local market are mainly imported. Therefore, these commodities are sensitive to changes in the real exchange rate.

6.5 Simulation Design, Assumptions and Results

Policy analysis with UgAGE requires two simulations. The first is the baseline. This models the growth of the Ugandan economy under business-as-usual assumptions for key drivers of growth. Included in the baseline for this paper is the assumption of normal rates in rainfall and temperature over the entire course of the simulation period. The baseline simulation incorporates observed national accounting data for the main macro variables[16] and forecasts adopted from the International Monetary Fund (IMF) for demographic variables determining employment and population growth.

The second simulation is the policy simulation which incorporates all the features of the baseline simulation plus shocks to agricultural output that reflect the impact of rainfall deficiencies associated with climate change. The results of the policy simulation are reported as deviations (percentage or otherwise) away from baseline values. We solve the model using GEMPACK (Harrison et al. 2016; Harrison and Pearson 1996).

The baseline simulation shows that, given the imposed growth rates for GDP at 6 per cent per annum and employment of 3 per cent, capital grows at 5 per cent per annum. For capital to grow at 5 per cent, annual investment must also grow by around 5 per cent. For all other macro variables, annual year-on-year growth rates vary between 5 and 6 per cent.

Throughout the simulation period, export growth exceeds imports. This is consistent with a fall in the terms of trade. *Ceteris paribus*, this requires real depreciation of the Ugandan Shilling. With no specific industry forecasts imposed during the baseline simulation, we find that all industries expand at rates similar to that of real GDP.

6.5.1 Modelling a Fall in Agricultural Output

It is assumed for the baseline that growth in agricultural production (in the absence of climate change) is roughly in line with GDP. Given initial values of output in the base year and year-on-year percentage changes over the simulation period, we are able to calculate the value of real output by sector in each year of the simulation period.[17]

In the policy simulation we reduce the output of agricultural sectors in line with the numbers shown in Table 6.3. To understand how we calibrate the output shocks, consider the case of coffee. According to the baseline, the real level of coffee bean output in 2050 is projected to be 5279 billion Shilling. According to Table 6.3, weather-related changes related to climate change will reduce coffee output in the long-run by 30 per cent. We assume that the long-run means 2050. Thus, we assume that the coffee output in the policy simulation in 2050 will be 30 per cent lower than in the baseline. In other words, in the policy simulation in 2050 the value of coffee production will be 3695 billion Shilling. Given an identical starting point, but different end points, we calculate for the policy simulation that production of coffee

[16]We imposed observed data for real investments, household and public spending and exports (UBOS 2017).

[17]The base year is 2010 where nominal and real value are similar. By multiplying the initial database value with the year-on-year percentage change in industry output, we are able to calculate the value of real output per year.

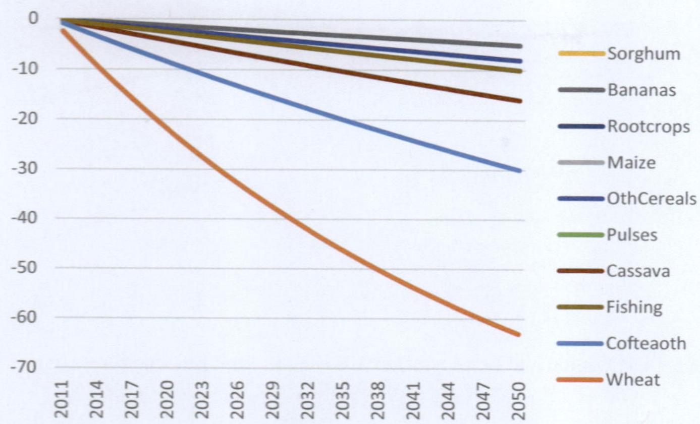

Fig. 6.3 Simulated fall in output levels of selected agricultural sectors (cumulative % changes)

production will grow at an average annual rate 0.89 percentage points less than in the baseline.

Figure 6.3 shows the cumulative percentage changes in output for the agricultural sectors, as implemented in the policy simulation. These changes in output are forced on the model, via model-determined changes in all-factor saving technological progress in each sector. For example, to achieve 62 per cent contraction in wheat production by 2050, the model determines year-by-year the amount of technological regress sufficient to contract wheat's supply schedule to achieve the assumed reduction in output.

6.5.2 *Overview of Macroeconomic Results*

Figures 6.4, 6.5, 6.6, 6.7 and 6.8 show results for the key macroeconomic variables, expressed as percentage deviations away from baseline values. For example, Fig. 6.4 shows that in 2050, climate change will reduce real GDP by 4.1 per cent relative to its level in the baseline forecast.

The headline result is that rainfall deficiency and temperature change associated with climate change will, to the extent that it harms agricultural production, suppress macroeconomic activity in Uganda.

We use a Back of the Envelope (BOTE) model to provide the reader with an understanding of the reasons why the model gives this macroeconomic outcome. The BOTE used here is identical to that described in more detail in Dixon and Rimmer (2002b).

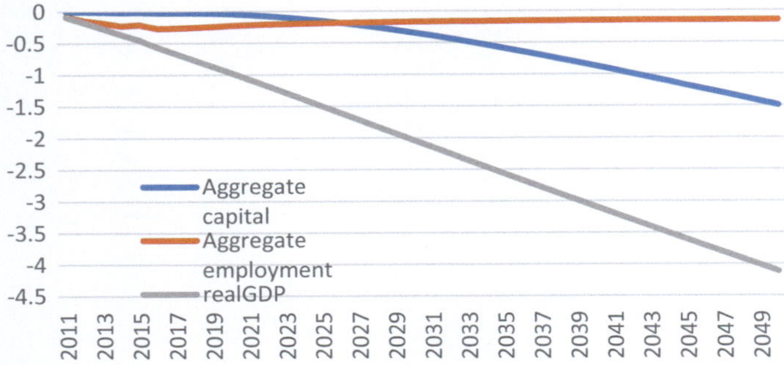

Fig. 6.4 Real GDP, Capital and Employment (% deviation from baseline forecasts)

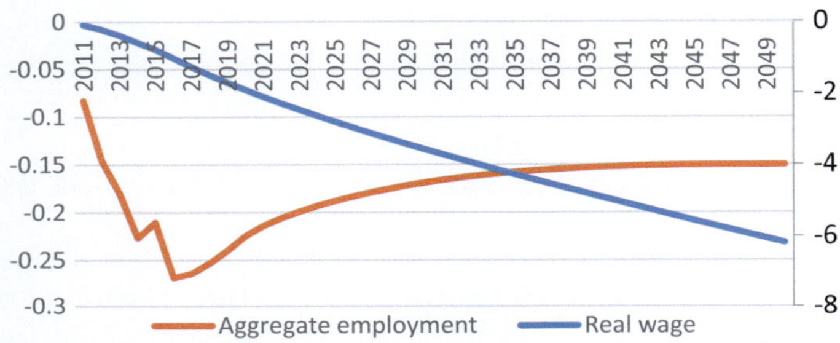

Fig. 6.5 Real wage rate and aggregate employment (% deviations from baseline)

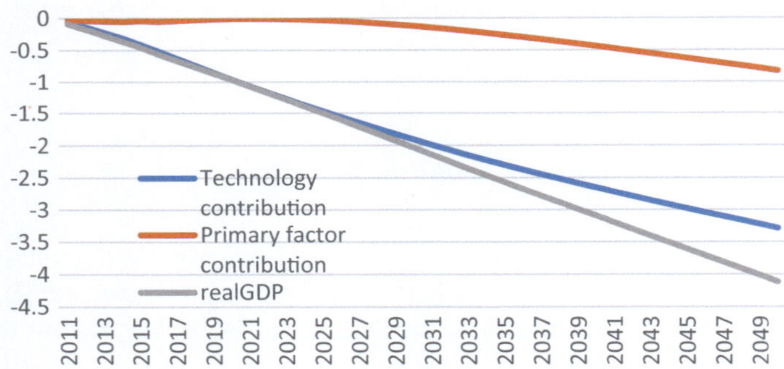

Fig. 6.6 Contributions to real GDP from the income side (% deviations from baseline)

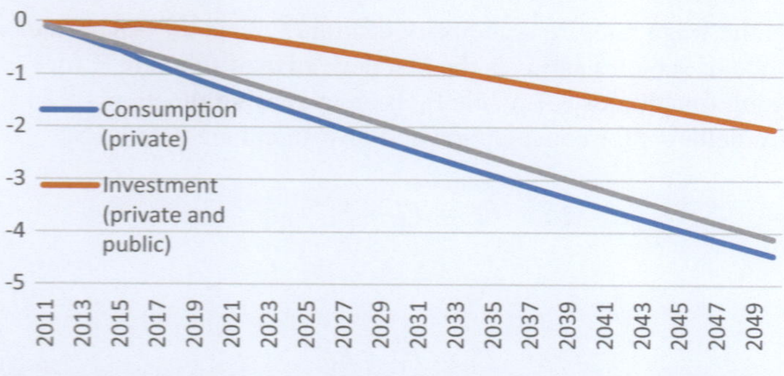

Fig. 6.7 Real investment, private consumption and GDP (% deviations from baseline)

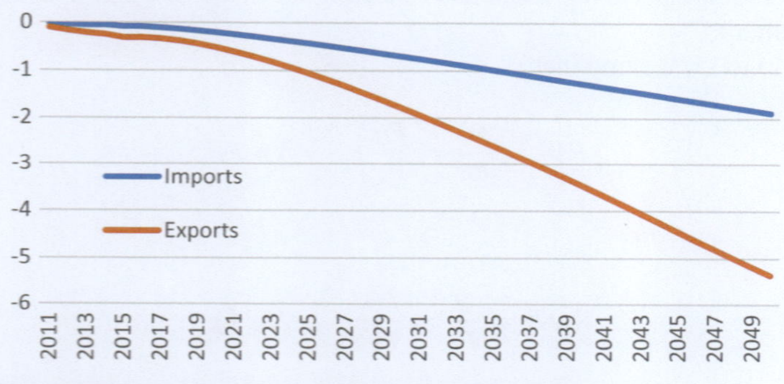

Fig. 6.8 Macroeconomic trade variables (% deviations from baseline)

6.5.3 A More Detailed Explanation of the Macroeconomic Results

Total production in the economy is a function of capital, labour and technology:

$$Y = A * F(K, L) \tag{6.1}$$

where: Y is domestic output; K and L are inputs of capital and labour; F is a homogeneous function of degree one (constant returns to scale); and A is a technology variable, with increases in A representing technological improvements.

Labour and capital are paid the value of their marginal products:

$$W = P_G * A * F_L \tag{6.2}$$

and

$$Q = P_V * A * F_K \tag{6.3}$$

where W is the wage rate; P_G is the price of grain[18]; Q is the rental rate on capital; P_V is the price of vehicles and F_i is the first derivative of function F, for i = K, L. F_K is a decreasing function of K/L while F_L is an increasing function.

The price deflators for consumption and investment are defined by

$$P_C = P_G^{\gamma} * P_V^{1-\gamma} \tag{6.4}$$

and

$$P_I = P_G^{\delta} * P_V^{1-\delta} \tag{6.5}$$

where γ and δ are positive parameters reflecting the share of grain in consumption and investment.

From (2) to (5) we find that

$$RW = \frac{W}{P_C} = \frac{P_G}{P_V}^{1-\gamma} * A * F_L \tag{6.6}$$

and

$$ROR = \frac{Q}{P_I} = \frac{P_G}{P_V}^{1-\delta} * A * F_K \tag{6.7}$$

where RW is the real wage and ROR is the real rate of return on capital. The price ration, $\frac{P_G}{P_V}$, can be thought of as the economy's terms of trade.

We start our explanation of results using Eq. (6.7). The reductions in agricultural production imposed in the modelling (Fig. 6.3), require significant deteriorations in the rates of technological progress within the agricultural sector. In terms of the BOTE model, this means a reduction to the technology variable (A). In the long run, the rate of return on capital (ROR) changes little relative to baseline levels due to capital adjustment.[19] Also, there is relatively little change in the terms of trade $\left(\frac{P_G}{P_V}\right)$. Thus, according to Eq. (6.7) there must be an increase in the marginal product of capital (F_K). That means the K/L ratio will fall, since F_K is a decreasing function of K/L. Figure 6.4 shows that the economy wide ratio of capital to labour does fall. In

[18] In this simple BOTE there is only one domestic industry, namely grain, and one imported commodity, namely vehicles. Consumers buy both goods while investors use both as inputs into capital production.

[19] In our modelling we assume that following a shock capital shifts between industries to eventually eliminate the initial disturbances in rates of return. For example, if the shock increases profitability in an industry, investment will rise (relative to baseline levels). The increase in investment will lead to increased capital. Over time, capital will continue to move into the industry until the initial disturbance in rate of return is eliminated. Note, that in the simulations reported here, the shocks are not once-off, but continue through the projection period. Thus, rates of return almost, but never fully, return to their baseline levels.

2050, there is little change in employment, but capital is down 1.5 per cent relative to its baseline level.

As F_K increases, so F_L must decrease, leading via Eq. (6.6) to a fall in the real wage rate (RW). As shown in Fig. 6.5, at the end of the period the real wage rate is down by more than 6 per cent relative to its baseline level.

The reduction in real wage induces industries to employ more labour. Accordingly, national employment moves back towards its baseline level (Fig. 6.5). Employment never converges fully because the contractions in agricultural output are not once-off, but are imposed progressively through the simulation period.

Equation (1) shows that with technological regress (a negative change to A) and reductions in real factor inputs (falls in K and L), real GDP (Y) must fall. The percentage change form of (1) is given by:

$$y = a + S_K * k + S_L * l \tag{6.8}$$

where y, a, k and l are the percentage changes in output, technology, capital and labour, and S_K and S_L are the shares of capital and labour in GDP. The average shares of capital and labour over the projection period are 0.61 and 0.29. Thus, according to Eq. (6.8),

$y = -3.31 - (0.61)*1.5 - (0.29)*0.15 = -4.27$.[20]

This is close to the long-run decrease in GDP shown in Fig. 6.4, while Fig. 6.6 confirms this effect on real GDP in the long run, as well as the fact that the decrease in technology is the driving force behind the drop in real GDP.

Figure 6.7 shows deviations in two of the main domestic expenditure components of real GDP, namely private consumption and investment (private plus public). In this simulation, public consumption is fixed to its baseline level. In each year, relative to real GDP real private consumption falls, while real investment rises. In our modelling, private consumption is determined by a consumption function that relates nominal consumption to nominal GDP (the proxy for household income) via a fixed marginal propensity to consume. Thus nominal household consumption changes in line with changes in nominal GDP. *Real* consumption, however, can deviate from *real* GDP due to changes in the price of consumption relative to the price of GDP. In this simulation, the price of consumption rises relative to the price of GDP because agricultural products (whose prices increase) have a greater representation in national consumption than in national value added.

Investment falls relative to baseline levels in line with the mild contraction in capital (Fig. 6.4)—because the economy requires less capital, it also requires less investment to maintain the required capital. However, investment falls by less than real GDP, in line with the relative changes in capital and GDP (see Fig. 6.3).

---

[20]GDP comprises the cost of capital, the cost of labour, the cost of natural resources and indirect taxes net of subsidies. The combined share of natural resources and indirect taxes is 0.10. In this simulation, the stock of natural resources is held fixed and indirect taxes changes little relative to baseline levels.

Overall, real GDP (Y) falls by more than real Gross National Expenditure (C + I + G). Hence the net volume of trade (X-M) must deteriorate. Figure 6.5 shows movements away from the baseline in the volumes of exports and imports. In 2050, the volume of exports is 5.3 per cent below its baseline level, while the volume of imports is down 2.0 per cent. The deterioration in net volume of trade is accompanied by real appreciation of the Ugandan currency. Real appreciation reduces the competiveness of Ugandan products exported to world markets and competing against imports on local markets.

In UgAGE, it is assumed that Ugandan producers face downward sloping world demand schedules for their products. In terms of a supply and demand diagram, the real appreciation causes the supply schedules for Ugandan products to shift to the left. This reduces export volumes, but also lifts the world prices at which they are sold. Thus, despite all the bad news, there is some good news with the Ugandan economy projected to experience a small improvement in its terms of trade as a result of the climate-change related contractions in agricultural output.

6.5.4 Industry Results

The UgAGE model produces results for 95 industries. In Fig. 6.9 we show the ten industries that are affected most by the shocks to agricultural production. For each of the industries we show the percentage deviation in total production in the final year of the simulation (2050) as a black bar. Four of the industries shown in Fig. 6.9 experience increases in production (relative to baseline levels)—the textiles sector ("TextilePrd"), the fabricated metal products industry ("FabMetalPrd"), the Annuals product industries[21] ("Annuals"), and the wearing apparel sector ("WearAppPrd"). The remaining six industries are projected to experience relatively large contractions (relative to baseline levels) in output in 2050.

For each of the ten industries we decompose the overall change in production into three contributing factors: (i) the local market effect, which shows the contribution of broad changes in the size of the local market; (ii) the domestic share effect, which shows the extent to which domestic activity is replaced by imports, and (iii) the export effect, which shows the contribution make by changes in exports.[22]

6.5.4.1 Textile Products, Rank 1 and Fabricated Metal Products, Rank 2

The biggest winners shown in Fig. 6.9 do not use agricultural products. Although they lose very small parts of their local markets (as shown by the red bars shown in

[21] Tobacco and Cotton.

[22] This decomposition is referred to as the Fan decomposition, in recognition of the Chinese scholar who developed the idea while at the Centre of Policy Studies.

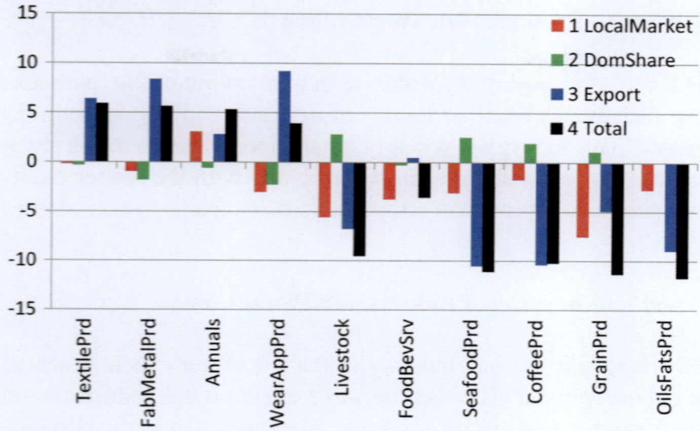

Fig. 6.9 Decomposition of the production changes for the main winning and losing industries in 2050

Fig. 6.9), exacerbated by increased imports that replace some domestic production (green bars), the industries gain as a result of increased exports (blue bars).

While there is a fall in total exports (see Fig. 6.8), some industries experience increased exports as a result of declining costs of production. For the textile and metals industries, their production costs are driven primarily by wages, which fall significantly (see Fig. 6.5).

6.5.4.2 Annuals Products, Rank 3

This sector has two sources of positive growth in the simulations: (i) it benefits from its intermediate linkages to textile products as an intermediate input; and (ii) increased export sales due to the fall in real cost of labour.

6.5.4.3 Wearing and Apparel Products, Rank 4

Wearing and Apparel products are significant exporters of commodities in the Ugandan context, and Fig. 6.9 shows that increased exports is a strong driver of the increase in total output, despite decreased local activity (drops in household and intermediate demands) and some import replacement of local production.

6.5.4.4 Livestock Products, 6th Worst Loser

Two strong forces are causing the decrease in total output of the livestock industry: (i) a decline (relative to baseline levels) in demand; and (ii) a strong decrease in foreign demand. This industry is strongly reliant upon maize, cassava and root crops for stock-feed. It is also a significant exporter, and with the higher input costs the industry becomes less competitive on world markets, leading to a fall in exports.

6.5.4.5 Food and Beverage Products, 5th Worst Loser

Food and Beverages is the one industry in the list of ten which is not affected by changes in export demand. The total negative effect on this industry is driven by a decrease in domestic demand for food and beverages. Domestic demand includes the demand from industries which use food and beverages as an intermediate input and household demand which use food and beverages as a final commodity. The fall in domestic demand is mostly driven household demand, which constitutes almost 90 per cent of local demand.

The local market demands less food and beverages because its output price rises as a result of higher input costs. The industry is strongly linked to the whole agricultural sector which is shocked in the model simulations.

6.5.4.6 Seafood Products, 4th Worst Loser

This industry buys all of its inputs from the fishing industry, which is directly shocked in the model simulations. Its inputs become more expensive, causing a decrease in exports of the commodity, which mostly explain the total decrease in its output.

6.5.4.7 Coffee Products, 3rd Worst Loser

Coffee products are similar to seafood: it has one large intermediate input, namely raw coffee beans, which are directly affected by the shocks. Coffee production costs increase with an adverse effect on exports, which drive total output completely.

6.5.4.8 Grain Products, 2nd Worst Loser

Grain products use large proportions of maize, cassava, sorghum and other cereals as inputs into its production process, all of whom are adversely affected by the climate change simulations that we have run. The decomposition in Fig. 6.9 shows two sources of deterioration in this industry: (i) local demand decreases, and this comes mostly from the livestock and food industries, and (ii) exports decrease as a result of higher output prices.

6.5.4.9 Oils and Fats Products, Biggest Loser

Oils and fats sells to most other industries as intermediate input, but mostly to the food and beverage industry, which we have seen decreases by 3.5 per cent over the forecast period, relative to the baseline. The main factor in its decline (relative to baseline levels), however, is again exports, which decline by 9 per cent.

6.6 Conclusion

Environmental and agricultural economists have forecast severe decreases due to climate change in Ugandan production of coffee, wheat, maize, sorghum, other cereals, pulses, root crops and fishing. The effects of these decreases have been simulated using the UgAGE model. The modelling shows that food production will decrease by 3.3 per cent relative to baseline levels in 2050. The shocked agricultural crops form about 12 per cent of total output, and their average decline is forecast to be 16.5 per cent. That would give a direct effect of 2 per cent decrease in total output (0.12*16.5). The model results show a decrease of 4.12 in real GDP, which is easily explained if we add the various indirect changes that result from the direct shocks, such as livestock (−8.6%), oils and fats products (−10.5%), seafood products (−9.8%), grain products (−10.3%), and coffee products (−9.3%), to name only a few.

In the introduction to the chapter we saw that food security consists of four elements, namely availability, access, utility and stability. The literature reports that many Ugandans are constantly without enough food to eat, throughout the country. If their population grow at 3 per cent per annum in the long-run, while agricultural production declines severely as a result of climate change, resulting in a decline in food production, then the food security position of Uganda would certainly worsen.

At least two portions of potential good news comes from the model: (i) even though the ten agricultural crops decreased in production between 5 and 63 per cent below the baseline values, total food and beverage products only decreased by 3.3 per cent below the baseline, and (ii) the production of annuals, textile products and wearing apparel on the one hand, and fabricated metal products on the other, increased.

The model showed that the adverse weather conditions in the wheat industry would cause almost a total switch to imports of wheat as input into the production of grain and other products. This addresses the availability of food, to some extent. Local droughts do not necessarily mean that Uganda could not have food available; they should always be able to import food. However, access to food includes the ability to afford and buy the food, and the ability to deliver imported food to all the regions of Uganda. Transport infrastructure in Uganda is underdeveloped, and even if the CGE model predicts increased imports of foodstuffs, without proper infrastructure this would remain a theoretical solution to the access problem. With real GDP decreasing more than four percentage points below the baseline forecast, the country is moving backwards and its citizens become worse off in the longer run, and less able to afford enough food.

The second point of potential good news points, in our view, to a possible solution for the food security problem in Uganda. Annuals have been forecast by the literature not to be affected adversely by climate change. Our modelling results show a strong switch towards the production of annuals, resulting in expansions in the textile and wearing and apparel sectors. We also saw increases in fabricated metals production. Factors of production move out of the affected industries and into other industries that fare better and are able to expand.

All of this provides an opportunity for the country and its policy makers. Uganda should rely less on agriculture and diversify its production activities to include other crops that are less vulnerable to climatic changes, and stimulate the manufacturing industry. However, this is much easier said than done. The manufacturing sector could hardly be developed with the current underdeveloped transport infrastructure in the landlocked Uganda. Manufacturing could only succeed with sufficient trade between regions, which is critically dependent on the said infrastructure.

What about the rain on Victoria Lake? We do not have irrigated agriculture as a separate industry in our model, and hence we did not include the amount of rain directly into the modelling simulations. Further research could be done by splitting the agricultural sector into irrigated agriculture and dryland agriculture, and specifically adding water supply and use data to the model. We could then repeat the simulations done but allow the model to switch from dryland to irrigate agriculture under conditions of severe drought.

Uganda has access to ample water from its lakes and rivers, but a very insignificant part of the country is connected to these water resources through pipes and formal infrastructure. Apart from the diversification of industry, the government would make the citizens much less vulnerable to climate change effects if the water and transport infrastructure are developed. All four aspects of food security would be addressed and improved in this case.

References

Adhikari U, Nejadhashemi AP, Woznicki SA (2015) Climate change and eastern Africa: a review of impact on major crops. Food Energy Secur 4:110–132

Allison EH, Perry A, Badjeck MC, Adger WN, Brown K, Conway D, Halls AS, Pilling GM, Reynolds JD, Andrew NL, Dulvy NK (2009) Vulnerability of national economies to the impacts of climate change on fisheries. Fish Fisheries 10:173–196

Asseng S, Foster I, Turner NC (2011) The impact of temperature variability on wheat yields. Glob Change Biol 17:997–1012

Boote KJ, Jones JW, Hoogenboom G, Pickering NB (1998) The CROPGRO model for grain legumes. In: Tsuji GY, Hoogenboom G, Thornton PK (eds) Understanding options for agricultural production. Academic Publishers, Dordrecht, pp 99–128

Broughton WJ, Nernandez G, Blair M, Beebe S, Gepts P, Vanderleyden J (2003) Beans (Phaseolus spp.)- model food legumes. Plant Soil 252:55–128

Brown L (2009) Plan B 4.0: mobilizing to save civilization. W.W. Norton and Company, New York

Bunn C, Läderach P, Rivera OO, Kirschke D (2015) A bitter cup: climate change profile of global production of Arabica and Robusta coffee. Clim Change 129:89–101

Calberto G, Staver C, Siles P (2015) An assessment of global banana production and suitability under climate change scenarios. In: FAO Climate change and food systems: global assessments and implications for food security and trade. FAO, pp 265–291

Dixon PB, Rimmer MT (2002a) Dynamic general equilibrium modelling for forecasting and policy: a practical guide and documentation of MONASH. North-Holland, Amsterdam
Dixon PB, Rimmer MT (2002b) Explaining a dynamic CGE simulation with a trade-focused BOTE analysis: the effects of eCommerce on Australia. Centre of Policy Studies Working Paper Series, G-136
Dixon PB, Parmenter BR, Sutton J, Vincent DP (1982) ORANI: a multisectoral model of the Australian economy. North-Holland, Amsterdam
FAO (2002) The state of food insecurity 2001. Rome
FAO (2003) Trade reforms and food security—conceptualizing the linkages. Rome
FAO (2013a) FAOSTAT online statistical service. FAO, Rome
Fermont AM (2009) Cassava and soil fertility in intensifying smallholder farming systems of East Africa. PhD dissertation. Wageningen University, Netherlands
Fisher G (2009) World food and agriculture to 2030/50: how do climate change and bioenergy alter the long-term outlook for food, agriculture and resource availability? http://www.fao.org/3/a-ak972e.pdf accessed 24 September 2018
Galvin KA, Thornton PK, Boone RB, Sunderland J (2004) Climate variability and impacts on east African livestock herders: the Maasai of Ngorongoro Conservation Area. Afr J Range Forage Sc 21:1–11
Harrison WJ, Horridge JM, Jerie M, Pearson KR (2016) GEMPACK manual, GEMPACK Software. http://www.copsmodels.com/gpmanual.htm accessed 24 September 2018
Harrison WJ, Pearson KR (1996) Computing solutions for large general equilibrium models using GEMPACK. Comput Ec 9:83–127
Hepworth N (2010) Climate change vulnerability and adaptation preparedness in Uganda. https://ke.boell.org/sites/default/files/uganda_climate_change_adaptation_preparedness.pdf accessed 24 September 2018
International Trade Centre (ITC) (2011) Cotton and climate change: impacts and options to mitigate and adapt. http://www.intracen.org/Cotton-and-Climate-Change-Impacts-and-options-to-mitigate-and-adapt/ accessed 24 September 2018
Jarvis A, Ramirez-Villegas J, Campo BVH, Navarro-Racines C (2012) Is Cassava the answer to African climate change adaptation? Trop Plant Bio 5:9–29
Liu J, Fritz S, Van Wesenbeeck CFA, Fuchs M, You L, Obersteiner M (2008) A spatially explicit assessment of current and future hotspots of hunger in sub-Saharan Africa in the context of global change. Glob Planet Change 64:222–235
Lobell DB, Field CB (2007) Global scale climate-crop yield relationships and the impacts of recent warming. Env Research Letters 2
Lobell DB, Burke MB, Tebaldi C, Mastrandrea MD, Falcon WP, Naylor RL (2008) Prioritizing climate change adaptation needs for food security in 2030. Science 319:607–610
Lobell DV, Banziger M, Magorokoso C, Vivek B (2011) Nonlinear heat effects on African maize as evidenced by historical yield trials. Nat Climate Change 1:42–45
Moat J, Williams J, Baena S, Wilkinson T, Gole TW, Challa ZK, Demissew S, Davis AP (2017) Resilience potential of the Ethiopian coffee sector under climate change. Nat Plants 3:1–14
Nakibuuka B (2016) Eliminating the culture of open defecation in Karamoja. http://www.monitor.co.ug/artsculture/Reviews/Eliminating-culture-open-defecation-Karamoja/691232-3021298-154wsqgz/index.html accessed 24 September 2018
Nelson GC, Rosegrant MW, Koo J, Robertson R, Sulser T, Zhu T, Ringler C, Msangi S, Palazzo A, Batka M, Magalhaes M, Valmonte-Santos R, Ewing M, Lee D (2009) Impact on agriculture and costs of adaptation. http://www.fao.org/fileadmin/user_upload/rome2007/docs/Impact_on_Agriculture_and_Costs_of_Adaptation.pdf accessed 24 September 2018
Ringler C, Zhu T, Cai X, Koo J, Wang D (2010) Climate change impacts on food security in sub-Saharan Africa: insights from comprehensive climate change scenarios. http://www.ifpri.org/publication/climate-change-impacts-food-security-sub-saharan-africa accessed 24 September 2018

Roos EL, Adams PD, van Heerden JH (2015) Constructing a CGE database using GEMPACK for an African country. Comput Econ 46:495–518

Tatsumi K, Yamashiki Y, Valmir da Silva R, Takara K, Matsuoka Y, Takahashi K (2011) Estimation of potential changes in cereals production under climate change scenarios. Hydro Process 25:2715–2725

Thornton PK, Jones PG, Alagarswamy G, Andresen J (2009) Spatial variation of crop yield response to climate change in East Africa. Glob Env Change 19:54–65

Uganda Bureau of Statistics (UBOS) (2017) Annual GDP data. Available at http://www.ubos.org/statistics/macro-economic/annual-gdp/ accessed 31 January 2017

Van Asten PJA, Fermont AM, Taulya G (2011) Drought is a major yield loss factor for rainfed East African highland banana. Agri Water Manage 98:541–552

Wortmann CS, Kirby RA, Eledu CA, Allen DJ (1998) Atlas of common bean production in Africa. CIAT-Pan-African Bean Research Alliance, Kampala

You L, Rosegrant MW, Fang C, Wood S (2005) Impact of global warming on Chinese wheat productivity. IFPRI, Washington, DC

Prof. Jan van Heerden is one of the pioneers of CGE modelling in South Africa and has been working on joint projects with members of CoPS in Melbourne for two decades. One of these projects was to build a national database and dynamic national CGE model for the Ministry of Finance in Uganda, and to train a group of twenty economists from various departments to use the model for policy analysis. A recently completed project was to build a regional database and model (TERM) for the Gauteng Provincial Treasury in South Africa, and train fifteen economists from two departments to use the model for policy analysis. In his personal research Jan mostly applies CGE models to do environmental policy analysis.

Dr. Louise Roos is a Senior Research Fellow at the Centre of Policy Studies (CoPS) at Victoria University. Prior to joining CoPS in 2011 she worked at the University of Pretoria (2000–2006) as a lecturer. Dr Roos holds a PhD from Monash University (2012) and a Master in Economics from the University of Pretoria, South Africa (2003). Louise's interests include the development and application of large multi-sectoral and multi-regional dynamic CGE models. She has developed models, including detailed multi-sector databases, of the South African, Ugandan, Saudi Arabian and Jordanian economies.

Prof. Philip Adams is an expert in applying and building CGE models for practical policy analysis. Prior to his current position at the Centre of Policy Studies (CoPS), Victoria University, Philip was Director and Professor at CoPS, Monash University (2004–2013). Philip was elected a Fellow of the Academy of the Social Sciences in Australia in 2016. In that year he was also awarded the GTAP Research Fellow distinction for the term of 2016 to 2019. Since completing his Ph.D., he has been involved in the implementation of several large models of the Australian economy. He has also been active in developing models for Saudi Arabia, Oman, Jordan, Uganda, South Africa, Taiwan, Denmark, and Thailand. Philip has around 60 refereed publications, published in a wide range of journals, including: Nature, the Journal of Policy Modelling and the International Journal of Forecasting. He is the co-author of a chapter on Environmental modelling in the *Handbook on CGE modelling* (2013, Elsevier B.V).

Dr. Nicholas Kilimani is a Lecturer in the Department of Policy and Development Economics, College of Business and Management Sciences, Makerere University in Kampala-Uganda. His research interests are in environmental and development economics. His publications including modelling the impact of drought induced productivity losses, the policy relevance of water resource accounting and the use of fiscal policy in the management and use of environmental resources.

Chapter 7
The Expansion of Irrigated Agriculture in Brazil and Potential Regional Limitations

Angel dos Santos Fachinelli Ferrarini, Joaquim Bento de Souza Ferreira Filho, Santiago Vianna Cuadra, Daniel de Castro Victoria and Mark Horridge

Abstract Regional expansion of irrigated agriculture may minimize risks for the farmer and strengthen food security for future generations. However, at the same time, it has a damaging effect on water availability and may worsen episodes of micro regional water scarcity. In this paper, we analyze scenarios of irrigation expansion in the Brazilian agriculture, using a computable general equilibrium model (the TERM-BR model) to assess the impacts on water demand for 2025. Simulations were carried out for areas deemed potentially suitable for irrigation based on National Water Resources Plan (PNRH) and Ministry of National Integration report (MI). We increase the regional detail to focus on some of the Brazilian Northeastern States, in order to assess the balance between regional water supply (as determined by the Climatic Water Balance) and demand. The comparative results of the Climatic Water Balance and the TERM-BR model for the region point to potential water availability problems in specific states, such as Alagoas and Pernambuco.

The original version of this chapter was revised: Authors' sequence has been changed. The correction to this chapter is available at https://doi.org/10.1007/978-981-13-6101-2_10

A. dos S. F. Ferrarini
Universidade Estadual de Maringa (State University of Maringa), Maringá, PR, Brazil
e-mail: angel.fachinelli@gmail.com

J. B. de S. Ferreira Filho (✉)
Department of Economics, Business and Sociology, School of Agriculture "Luiz de Queiroz", University of Sao Paulo, Piracicaba, SP, Brazil
e-mail: jbsferre@usp.br

S. V. Cuadra · D. de C. Victoria
Empresa Brasileira de Pesquisa Agropecuária (Brazilian Agricultural Research Corporation), Campinas, SP, Brazil
e-mail: santiago.cuadra@embrapa.br

D. de C. Victoria
e-mail: daniel.victoria@embrapa.br

M. Horridge
Centre of Policy Studies, Victoria University, Melbourne, VIC, Australia
e-mail: mark.horridge@gmail.com

G. Wittwer (ed.), *Economy-Wide Modeling of Water at Regional and Global Scales*, Advances in Applied General Equilibrium Modeling,
https://doi.org/10.1007/978-981-13-6101-2_7

Keywords Brazil · Water demand and supply · Irrigation expansion · Regional water scarcity · CGE models

7.1 Introduction

The Brazilian National Water Resources Plan constitutes a strategic set of inter-institutional actions aiming the improvement of water supply and demand in the country, in terms of both quality and quantity (MMA 2006a). The plan considers 12 disaggregated hydrographic regions in 56 planning units; it also takes into account regionalization in Special Planning Situations (SPS), which constitute territorial spaces whose limits need not necessarily coincide with those of a water basin (MMA 2006b). Therefore, the plan takes into consideration the physical, biotic, cultural and socio-economic diversity of Brazilian hydrographic regions with regional and national integrations, as shown in Fig. 7.1.

The Ministry of National Integration (MI) report on irrigated agriculture states that the new public irrigation projects planned and implemented with public resources should be in accordance with the territorial analysis of the country. This study shows that the total additional irrigable area in Brazil would amount to 75 million hectares (Mha), out of which 14 Mha (18%) are unlikely to expand irrigation for physical and

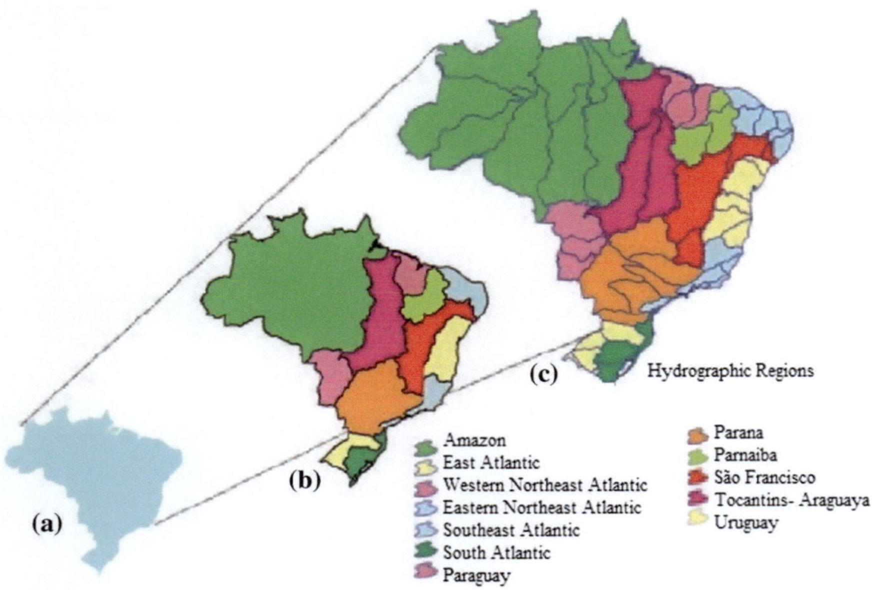

Fig. 7.1 PNRH aggregation levels, **a** Brazil, **b** National Hydrographic Division (Sub-1_), **c** 56 planning units. *Source* MMA (2006b)

institutional constraints, and 34 Mha (45%) would require public intervention and incentives (MI 2014).

The PNRH shed light on the need to intensify researches on water usage in all regions of the country. The northeastern semiarid region is prominent in this sense, since it is a region endowed with great biological diversity, with many species of plants, legumes and fruits are in areas susceptible to desertification. In this area, however, which comprehends the Brazilian semiarid and its surroundings, more than 80% of households do not have water supply and sewerage services.

The worldwide increase in demand for food influences the practice of irrigation, as one important item for agricultural productivity increases. Irrigated areas have increased in Brazil during the last two decades, reaching 10% of total annual crops area in 2006, as can be seen in Table 7.1.

Annual crops including sugarcane, rice, corn and soybeans have the largest irrigated areas in the country. Among permanent crops, fruit production is prominent in irrigation.

The Brazilian Agricultural Census of 2006 in IBGE (2009) reported that the Northeast region presented more than 1 million irrigated hectares in 2006. At the same time, as result of water availability reduction caused by a plurennial drought from 2012 to 2015, conflicts motivated by water usage in the region have been on the rise over the years (Martins and Magalhães 2015).

The main reasons ranged from the threat of expropriation, non-fulfillment of legal procedures, destruction and/or pollution to access hindrance caused by clandestine barriers and water stream deviation, among others.

Table 7.2 shows the evolution in the number of conflicts motivated by water use in the states of the Northeast region, as well as the number of families involved, which reached 14,518 families in 2010.

Northeast Brazil suffers a disproportionate number of conflicts. Besides natural phenomena, like a dry year, changes in Brazil's economic conditions can potentially influence its frequency and severity. Increases in domestic and export demands for agricultural products may elevate the rise in the number of conflicts, due to increases in water usage demand (MMA 2006c). In this paper, a computable general equilib-

Table 7.1 Total crops area and irrigated area in Brazil, 2006

	Annual crops			Permanent crops		
	Area (ha)	Irrigated area	%	Area (ha)	Irrigated area	%
North	1,837,143	103,945	6	441,563	24,870	6
Northeast	12,092,757	1,302,767	11	1,595,151	197,517	12
South	16,343,394	1,576,337	10	406,255	25,317	6
Midwest	12,264,580	777,796	6	83,050	16,679	20
Southeast	7,737,673	1,381,213	18	2,520,623	509,691	20
Brazil total	50,275,547	5,142,058	10	5,046,642	774,074	15

Source Brazilian Agricultural Census of 2006 at IBGE (2009)

Table 7.2 Conflicts motivated by water usage in the states of Brazil's Northeast region from 2006 to 2015

Regions	Number of conflicts									
	2006	2007	2008	2009	2010	2011	2012	2013	2014	2015
Alagoas (AL)		2				3		2	2	1
Bahia (BA)	3	2	7	2	15	9	8	21	26	27
Ceará (CE)	2	4	6	5	8	2	4	4	1	2
Maranhão (MA)		2	1	3	8	4	8		4	5
Pernambuco (PE)	3	7	2	2	5	11	5	5	8	9
Piauí (PI)		3	1	3		1	1			2
Paraíba (PB)	1		2	1		2	2	3		
Rio Grande do Norte (RN)	2	4	2		2	1			1	
Sergipe (SE)	2							1		
Northeast conflicts	13	24	21	16	38	33	28	36	42	46
Brazil total	45	87	46	45	87	68	79	93	127	135
% of Brazil's NE	29	27	46	35	44	48	35	38	33	34
Families involved ('000s)	1.2	9.8	8.2	4.4	14.5	8.6	3.9	4.2	10.4	9.3

Source CPT (2016)

rium model of Brazil, the TERM-BR model, is used to simulate scenarios of irrigation expansion proposed by the National Water Resources Plan (PNRH), as described in MMA (2006a), which lists areas potentially suitable for irrigation reported by the MI (2014). With the aid of the TERM-BR model, we estimated the implied water demand increase in the PNRH scenarios until 2025. These demand data are compared to water supply data, estimated using the Climatic Water Balance (CWB), for Northeast states in Brazil, to infer about the water balance in the region.

This study contributes to the water use literature in Brazil in several aspects. First, we elaborate an original water use database, by economic activity and by region. Second, to the best of our knowledge this is the first time the PNRH scenarios are simulated in a broad economic model, with regional details. Third, the comparison between results for water demand with a regional detail, estimated with detailed physical regional information is a methodological and empirical advance as well. We hope this paper will contribute to the future expansion of this methodology for the whole country.

7.2 Methodology

The methodology of this study comprised three main different steps: the elaboration of a detailed water use database, the use of a CGE model to estimate water demand, and the use of a Climate Water Balance method to estimate water supply. In this section, we describe the main feature of those models.

The TERM-BR model is an interregional, bottom-up, dynamic computable general equilibrium model (CGE) of Brazil, based on the theoretical structure of the Australian TERM model, calibrated for 2005. The model includes annual recursive dynamics and a regional representation, distinguishing 15 Brazilian regions, 36 sectors plus final demand. The complete structure of the TERM model database is detailed in Horridge (2012) and versions adapted for Brazil may be consulted in the studies of dos Santos (2006), Fachinello (2008), Fachinello and Ferreira Filho (2010), Ferreira Filho and Horridge (2014), among others.

The evaluation of water usage in the TERM-BR model separates agricultural lands into irrigated agriculture and dry farming land. The increase of regional agricultural production depends on both the growth of areas (irrigated and non-irrigated) and productivity. Land productivity, in turn, depends on the irrigated area, increasing with irrigation. The following equations describe the relation between the increase in irrigation and productivity in agriculture. Initially, we note that total land productivity (K) in agriculture is a weighted average of the productivity in irrigated and non-irrigated land:

$$\mathrm{K} = \mathrm{SHR}i \cdot \mathrm{K}i + \mathrm{SHR}n \cdot \mathrm{K}n \qquad (7.1)$$

in which K is total land productivity, Ki is irrigated land productivity and Kn is non-irrigated land productivity.

Equation (7.1) shows that total land productivity for a particular crop in a region is a weighed average of irrigated (SHRi) and non-irrigated (SHRn) land productivities. Expanding the irrigated area expands both water use and productivity, since irrigated land productivity is higher than non-irrigated land. The average productivity variation would be given by:

$$d\mathrm{K} = \mathrm{K}i \cdot d\mathrm{SHR}i + \mathrm{K}n \cdot d\mathrm{SHR}n \qquad (7.2)$$

$$d\mathrm{K} = \mathrm{K}i \cdot \mathrm{SHR}i \cdot shrig + \mathrm{K}n \cdot \mathrm{SHR}n \cdot shrnig \qquad (7.3)$$

Equation (7.3) shows the variation in the average productivity (dK = k*) as the weighted average of variation in irrigated and non-irrigated areas, where *shrig* and *shrnig* are the percent variation in the respective shares. If we define x as the relative land productivity between both types of land, and Ki > Kn, it follows that:

$$\mathrm{Kn/Ki} = \mathrm{x}\,, \quad 0 < x < 1 \qquad (7.4)$$

Therefore, making all necessary substitutions and differentiations, Eq. (7.5) shows the elasticity of productivity in relation to irrigated land:

$$\frac{\partial k^*}{\partial \text{shrig}} = \frac{\text{SHRi } (1 - \text{x})}{(1 - \text{x})\text{SHRi} + \text{x}} \tag{7.5}$$

The irrigated area share variation (*shrig*) is an exogenous variable in the model, determined by the economic policy scenarios.

7.2.1 The Water Usage Database

The building of the model water database involved different data sources. The main source of information were the technical coefficients of water use from the Ministry of Environment of Brazil MMA (2011).[1] Other sources like the Annual Industrial Research (PIA) at a product level, the Agricultural Census of 2006, the National Research for Household Sampling (PNAD) and National Research for Basic Sanitation (PNSB) were also used. In some cases, field interviews were required, in companies belonging to specific sectors.[2]

The initial agricultural productivity database for irrigated and non-irrigated activities was prepared through an extensive literature survey, which detailed description can be seen Ferrarini (2017). Data for water usage (consumption) in millions of cubic meters (Mm^3) can be seen in Table 7.3 which shows aggregated sector data for Brazil and selected states, located in the Northeast region, the focus of our analysis.

7.2.2 The Climate Water Balance and the Brazilian Northeast

Climate data from the CRU (Climate Research Unit, version 3.2) was used to estimate the water balance in the Northeast region. The Thornthwaite and Mather method (1955) was used to derive the monthly Climatic Water Balance (CWB) and may be consulted, for instance, in Doorenbos and Kassam (1994), Amorim Neto (1989), Pereira (2005) and Varejão-Silva (2006). Figure 7.2 shows the states located in the Brazilian Northeast region (panel A), as well as the water basins inside that region and their linkage to basins coming from other regions (panel B), which were used in the CWB water surplus calculations.

The water demand database was structured at state or municipal level. The annual average water surplus estimated from CWB was aggregated into larger units

[1]This report brings technical coefficients of water consumption, withdrawal and return for agriculture (municipal level) and other economic activities (national level).

[2]Namely, Leather, Forestry, Furniture, Dairy for direct interviews. Mining and Oil Extraction coefficients were obtained from Sustainability Reports.

(hydro-regions) which takes into account watershed divisions and state boundaries (Fig. 7.2b). The blue circle illustrates the interaction between basins in the Northeast region and others located in states of the South and Southeast Brazil.

Table 7.3 Water use by activity and region

	Agriculture	Livestock farming	Industry	Services	Families	Total
Brazil	20,084	3216	15,863	515	2454	42,132
Maranhão (MA)	180	100	54	14	87	435
Piauí (PI)	100	73	19	6	40	238
Ceará (CE)	490	74	424	16	85	1089
RGNorte (RN)	496	25	81	7	37	646
Paraíba (PB)	631	36	146	8	41	862
Pernambuco (PE)	2191	64	162	22	102	2541
Alagoas (AL)	3719	18	65	6	32	3840
Sergipe (SE)	108	17	51	6	30	212
Bahia (BH)	1126	238	483	25	124	872

Millions of m^3, 2006. Brazil and selected states
Source MMA (2011)

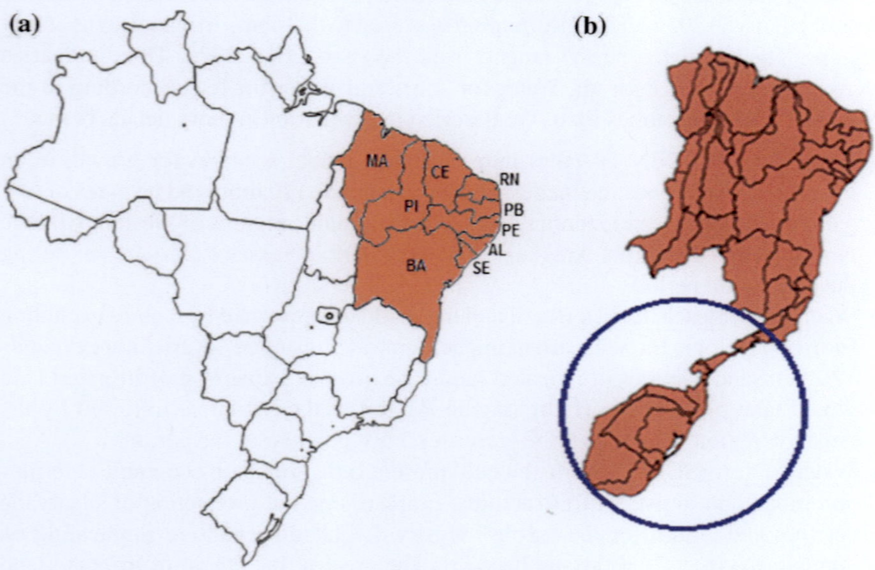

Fig. 7.2 **a** States in the Brazilian Northeast region; **b** Linkage between basin and states. *Source* Based on CWB data

The water supply at these hydro-regions were obtained from water surplus from CWB model minus water demand, for the hydro-region in question, obtained from the TERM-BR model. In the water availability calculations, water surpluses in each region (as calculated by water supply by CWB data minus water demand from the TERM-BR model) were transferred to hydro-region located immediately downstream. This process, once calculated for the entire regional area, allows a detailed accounting of water net availability in every basin, in different policy scenarios.

7.3 Scenarios and Model Closure

As mentioned before, the model initial database is for 2005. The first step for the simulation was to update it until 2012 with the observed macroeconomic, land use and population data, as well as for irrigated areas in Brazil.[3] Observed GDP evolution, for example, was imposed to the model in the historical period. After this period we started a forecast period, In this historical simulation, price of commodities increased annually 1% above the price of manufactured goods in the external market, and the GDP growth followed the National Water Resources Plan (PNRH) scenarios: GDP of 4.5% for scenario 1 (SCEN1), 3.5% in scenario 2 (SCEN2) and 1.5% in scenario 3 (SCEN3).

The scenarios were designed using information on areas potentially suitable for irrigation classified as being of maximum interest for public intervention (MIIP) as described in MI (2014). This information was used to distribute irrigated areas across regions. The simulated period ranged from the years 2013–2025. Three scenarios were analyzed: Water for all, Water for some and Water for few, according to the National Water Resources Plan. We describe the scenarios in more details below.

- Water for all (SCEN 1): takes into account a global scenario for growth in an integrated, continuous manner. Agriculture expands 170 thousand hectares of irrigated land, with greater increment in the hydrographic regions of Western Atlantic Northeast and Tocantins-Araguaia. Brazilian GDP increases 4.5% per year during the simulation period.
- Water for some (SCEN 2): Brazil and the world are governed by a more excluding form of development, with strong impacts on water resources. Agriculture expands 120 thousand hectares of irrigated land. The greatest expansion of irrigated land would take place in the Hydrographic Region of the Parnaíba, followed by the Amazon region. Brazilian GDP increases 3.5% per year in the period.
- Water for few (SCEN 3): Brazil would present little growth in economic activities and urban and logistic infrastructures. Emphasis on the production of sugarcane for fuel and cotton for the textile industry. Irrigated agriculture implicating an annual growth of 70 thousand hectares. The greatest increment of irrigated land would take place in the Hydrographic Region of the Parnaíba, followed by the Western Northeast Atlantic. Brazilian GDP increases 1.5% per year.

[3]The year 2012 is the last year for which information referring to irrigated land is available in Brazil.

In the policy closure, the share of irrigated land (shrig), by culture and region, is exogenous, and the target for the simulations.

7.4 Results

7.4.1 Evolution of Water Demand in Brazil According to the PNRH Scenarios

Results of the three simulated scenarios can be seen in Tables 7.4 and 7.5. In the first scenario (Water for all) the larger increase in irrigated land generates a higher increase in GDP, as expected. This is caused by the larger increases in agricultural productivity generated by increasing irrigated areas.

The larger expansion of irrigation in scenario 1 would cause a larger increase in agriculture productivity and GDP. This is accompanied by a larger water use, which is caused either by increasing use in agriculture and in the other sectors of the economy, through the multiplier effect. The increase in irrigated areas, however, is uneven in the territory. The allocation of irrigation increases is done through a mapping of areas potentially suitable for irrigation, as reported by the MI (2014). With this, the

Table 7.4 Scenario results. Macroeconomic variables (% deviation from baseline in 2025)

	Water for all	Water for some	Water for few
	SCEN1	SCEN2	SCEN3
Household consumption	0.080	0.056	0.032
Real investments	0.189	0.145	0.092
Export volume	0.145	0.109	0.063
Real wages	0.170	0.120	0.065
Real GDP	0.105	0.074	0.042

Table 7.5 Water use (Mm^3/year) and area of irrigated land increase (Mha/year)

Description	Water for all	Water for some	Water for few
	SCEN1	SCEN2	SCEN3
Water usage			
National water usage (Mm^3/year)	7554	5349	3007
National water usage (% change)	10.56	7.92	4.97
Irrigated area			
Change in irrigated area (Mha/year)	2.49	1.77	1.00

The updating of baseline areas after the PNRH report was published led to a slightly different total in projected irrigated area than the original scenarios

Table 7.6 Regional GDP increases (% deviation from baseline in 2025)

Regions (macro regions)	Regional GDP		
	SCEN 1	SCEN 2	SCEN 3
Rondonia (N)	0.13	0.08	0.04
Amazon (N)	0.23	0.18	0.11
ParaToc (N)	0.30	0.22	0.13
MarPiaui (NE)	0.10	0.07	0.04
PernAlag (NE)	0.03	0.02	0.01
Bahia (NE)	0.06	0.04	0.02
RestNE (NE)	0.03	0.02	0.01
MinasG (SE)	0.18	0.13	0.07
RioJEspS (SE)	0.01	0.01	0.00
SaoPaulo (SE)	0.03	0.02	0.01
Parana (S)	0.12	0.09	0.05
SCatRioS (S)	0.15	0.11	0.06
MtGrSul (CW)	0.20	0.15	0.08
MtGrosso (CW)	1.18	0.82	0.48
Central (CW)	0.16	0.11	0.06

N North region; *NE* Northeast; *SE* Southeast; *S* South, *CW* Center-west

increases in agricultural productivity, and the ensuing economic linkages effects, result in different benefits for different states, as shown in Table 7.6.

The state of Mato Grosso, in the center-west region of Brazil, would show the larger GDP increase, as result of the allocation of irrigation increases. This state is the third largest in Brazil, has the largest potential for irrigation expansion, with a high volume of fresh water, numerous rivers, aquifers, springs, and the planet's largest floodable area (called Pantanal). The state is also one of the most important grain producers in Brazil.

The crops with the larger potential for irrigation expansion would be those with the largest planted areas in Brazil, like rice, sugar cane and soybeans. Unlike rice and sugar cane, however, soybeans uses a low water volume in irrigation (Table 7.7). Sugar cane expansion stands out in all scenarios, in terms of water use. Sugar cane is used in Brazil for sugar and energy. Sugar cane ethanol is used both as a fuel, either directly or mandatorily mixed with gasoline. More recently, it has been used for electricity generation from bagasse as well. This agricultural activity would expand by 549,342 irrigated hectares in SCEN1, of which 18,892 irrigated hectares (3.43% of the national total) would be located in the Northeast region, the driest region in Brazil. The "Other crops" aggregate includes agricultural crops such as tomatoes, avocados, peanuts, potatoes, peas, onions, beans, and sunflower, among others. Many of those activities are essential for regional development and income generation, as well as for food security in Brazil.

Table 7.7 Modeled changes in land and water use (deviations from the baseline, 2025)

Crops	SCEN 1			SCEN 2			SCEN 3		
	Hectares	(million m^3)	Variation %	Hectares	(million m^3)	Variation %	Hectares	(million m^3)	Variation %
Rice	520,948	1808	29	379,394	1318	21	215,076	748	12
Maize	310,941	480	64	226,119	349	47	129,651	201	27
Wheat and cereals	11,465	13	45	8157	9	31	4461	5	18
Sugar cane	549,342	2940	19	380,084	2030	13	205,888	1097	7
Soybeans	442,251	654	96	303,086	437	63	176,411	254	36
Other crops	242,212	587	22	176,900	427	16	104,030	251	9
Cassava	48,626	95	58	35,505	69	42	20,617	40	24
Tobacco	1615	0	5	1181	0	4	672	0	2
Cotton	109,250	195	91	79,461	142	67	45,683	82	39
Citrus fruits	44,972	143	2	32,391	103	18	18,480	59	10
Coffee	61,891	170	18	43,478	120	12	24,926	69	7
Forestry products	124,631	442	121	92,057	327	90	54,503	194	53
Total	2,468,144	7528		1,757,813	5331		1,000,398	2998	

Irrigated areas (ha) and water use (Mm^3)

The results in Table 7.7 refer to a "full irrigation" regime, or a watering regime implying irrigation during all the culture's life cycle, and are generated from the PNRH projections. This may not be required in all times and regions. The table represents then an upper limit for water demand in irrigation. Still, the "potentially suitable area" criteria used by MI (2014), which was used as the base criterion for irrigated land expansion scenarios, does not mean that irrigation expansion would be readily available, once infrastructure investment would be required, in order to meet the expansions in water demand.

The results on water demand increases, however, are important indicators of the potential pressure on this natural resource arising from the PNRH scenarios. The feasibility of those scenarios, however, must be matched by water supply, which is a variable with a high regional specificity. The comparison of water demand and supply, however, is a complex issue, especially in a country with a large territory as Brazil, with different hydrographic characteristics, and in which the water basins can cross many states. For this purpose, in next section we focus in one specific region, the Northeast region of Brazil. As mentioned before, this is the driest region in the country, and for which expanding irrigation can potentially face water constraints.

7.4.2 Effects of the Increase in Irrigated Areas in Northeastern Brazil

Next, we focus on policy results from SCEN1 to deal with the northeast region in more detail. We start with the irrigated area expansion in the states inside the region (Table 7.8). The greater expansion occur in the states of Maranhão (MA), Piauí (PI) and Bahia (BA), which are the ones with the greater potential for expansion of irrigated areas.

We note that Alagoas, Pernambuco and Ceara states have a great climatic diversity and low regional water availability, which result in a lower amount of potentially irrigable areas. Data in Table 7.8 show that for those three states the expansion of irrigated areas would be smaller, but still impacting on water use. Rio Grande do Norte and Sergipe, on the other hand, have in sugar cane their main agricultural product. Even though small states both in extension and in irrigated area, being concentrated in sugar cane means a high water demand for irrigation, once the culture is the most demanding of all agricultural activities.

Bahia has the greatest diversity in agricultural production, has the most extensive coastline in the Atlantic Ocean, and is the one with the largest territorial extension, population and GDP in the region. Maranhão and Piaui, on the other hand, have large extensions of land in the Cerrados biome, with great potential for expansion of grains. These states also have the most important underground water reserves in the region. The results of the irrigated area expansion showed above on water use can be seen in Table 7.9.

Table 7.8 Modeled irrigated areas, by state in the Northeast region, 2025

Crops	MA	PI	CE	RN	PB	SE	PE	AL	BA	Total
Rice	11,473	24,755	75	25	26	83	89	58	1455	38,039
Maize grains	4604	4805	152	33	83	57	10	1	6757	16,502
Wheat and cereals	0	0	0	0	0	0	0	0	30	30
Sugar cane	9249	1391	47	490	504	299	72	96	6744	18,892
Soybeans	7254	0	3	0	0	0	0	0	10,177	17,434
Other crops	3911	11,493	157	77	54	64	200	10	24,149	40,115
Cassava	4320	636	1	2	1	10	31	31	1425	6456
Tobacco	0	0	0	0	1	1	0	4	84	90
Cotton	0	0	0	4	5	0	0	0	4700	4709
Citrus fruits	75	176	2	0	0	20	14	9	545	841
Coffee	0	0	0	0	0	0	1	0	6434	6435
Forestry prod-ucts	0	0	0	0	0	0	2	0	4051	4053
Hectares	40,887	43,255	437	632	674	533	419	209	66,551	153,596

Key MA (Maranhão), PI (Piauí), CE (Ceara), RN (Rio Grande do Norte), PB (Paraíba), SE (Sergipe), PE (Pernambuco), AL (Alagoas), BA (Bahia)

The results show, for example, that the expansion of 24,755 ha of rice in the state of Piauí (PI) (Table 7.8) would increase water use by 106.98 million cubic meters in 2025 (Table 7.9), which represents 4331 m^3 per hectare. Notice the strong increase in sugar cane in Maranhão and Bahia, and rice in Maranhão and Piauí.

Ceara is another important state in the region. However, rainfall distribution is rather heterogeneous in the state, and some regions present high evaporation and low precipitation and/or irregularity of rains, limiting access to water. The regional water availability has been an obstacle to the expansion of irrigated areas in the Northeast Brazil, and a key point in this study is to analyze the river flows in basins and sub-basins that supply the Northeastern states of Brazil, to be seen in what follows.

As mentioned before, the knowledge about how the water basins interact is essential for water supply estimates, since the watersheds normally cross over states boundaries: some states receive water from river basins that have catchment area in other regions. In other cases, the water catchment area can be in the same as that of the supply, which reduces micro regional and state water availability. This is, for example, the case of basins in Rio Grande do Norte (RN) and Paraíba (PB) states, displayed in Fig. 7.3.

Table 7.9 Water use in SCEN1 (million m^3 deviation from the baseline, 2025)

Crops	MA	PI	CE	RN	PB	SE	PE	AL	BA	Total
Rice	49.58	106.98	0.44	0.15	0.15	0.48	0.60	0.39	567	164.44
Maize Grains	15.50	16.17	0.52	0.11	0.28	0.19	0.03	0.00	9.92	42.72
Wheat and cereals	0.00	0.00	0.00	0.00	0.00	0.00	0.00	0.00	0.07	0.07
Sugar cane	73.25	11.01	0.49	5.14	5.29	3.14	0.65	0.86	73.15	172.98
Soybeans	12.19	0.00	0.00	0.00	0.00	0.00	0.00	0.00	22.23	34.42
Other crops	15.19	44.65	0.85	0.42	0.29	0.35	1.37	0.07	122.06	185.25
Cassava	23.09	3.40	0.00	0.01	0.00	0.04	0.15	0.15	5.14	31.98
Tobacco leaves	0.00	0.00	0.00	0.00	0.00	0.00	0.00	0.01	0.07	0.08
Cotton	0.00	0.00	0.00	0.03	0.03	0.00	0.00	0.00	16.87	16.93
Citrus fruits	0.35	0.82	0.03	0.00	0.00	0.33	0.10	0.06	2.75	4.44
Coffee beans	0.00	0.00	0.00	0.00	0.00	0.00	0.01	0.00	27.90	27.91
Forestry prod-ucts	0.00	0.00	0.00	0.00	0.00	0.00	0.02	0.00	12.20	12.22
Total	189.14	183.03	2.33	5.85	6.05	4.53	2.92	1.55	298.03	693.43

In Fig. 7.3, Panel A shows the flow of river basins inside northeast Brazil, by basin code, while Panel B exemplifies the flow of basins from other states to Sergipe.

An example will help understand data in Fig. 7.3. The state of Sergipe has catchment areas in Goiás state (basins 746 and 747), Minas Gerais[4] (basin 745), Bahia (basin 743) and in the Federal District[5] (basin 748). In order to know the water supply in Sergipe, then, the water surpluses transferred from those basins to basins 741 and 753 in Sergipe (Panel B) have to be computed, netting out the water consumption in each region upstream.

The same procedure described above was used for the other states in the region, for all basins, what allowed the computation of water supply by each state.[6] For the sake of consistency check, the results of water supply calculated here were compared with the averages supplied by the National Agency of Waters—ANA (2007). In most cases, the differences were in the range of 5%, although in the Parnaiba basin the difference was considerably higher.

---

[4]Minas Gerais state is located in southeastern Brazil.

[5]Federal District, Goiás are located in Midwest region.

[6]These calculations were performed with specific hydrology programs, and not in the CGE model.

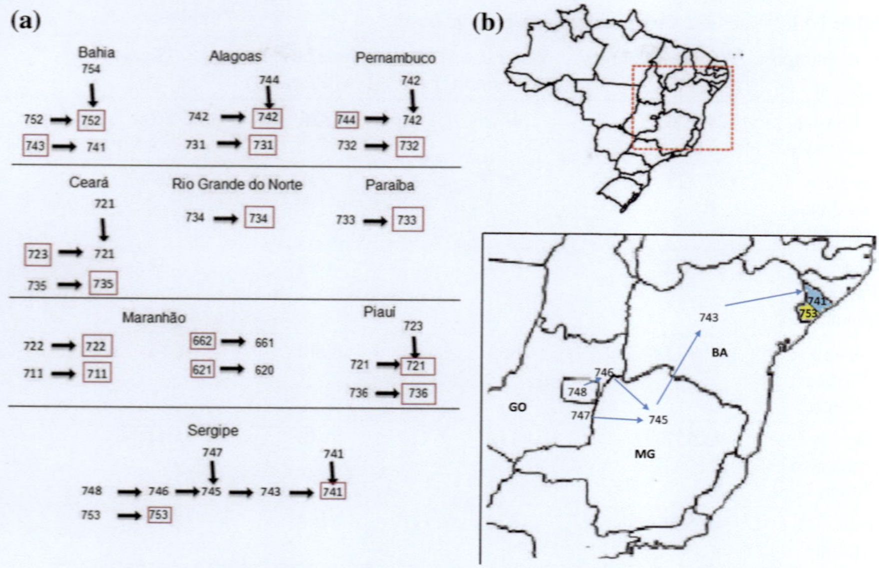

Fig. 7.3 Flowchart of the hydrographic basins in the Northeast regions of Brazil

Table 7.10 Modeled water supply and demand for the states in the Northeast region of Brazil

UF	2005 (in million m^3)			2025 in SCEN 1 (million m^3)		
	Supply (A)	Demand (B)	B/A (%)	Supply (A)	Demand (B)	B/A (%)
AL	8611	3864	45	8528	4277	50
PE	10,371	2605	25	11,060	2898	26
RN	3802	645	17	3802	782	21
PB	7306	800	11	7306	988	14
PI	41,260	247	1	41,177	476	1
MA	161,992	436	0	162,107	816	1
BA	39,297	2210	1	38,644	3141	8
SE	96,375	222	0	88,633	263	0
CE	56,240	891	2	18,597	1072	6

Supply, Demand (Mm^3) and ratio demand/supply. Comparison between 2005 (base) and 2025 (policy)

The results of the supply calculations, as well as the comparison with water demand data from the CGE model can be seen in Table 7.10, which shows, both for 2005 (initial year) and 2025 (our final simulation year), the balance between water supply and demand, and the demand/supply ratio.

The maximum demand/supply ratio would occur in Alagoas, where demand would reach 50% of total water supply in Scenario 1. Notice that the increase from 2005 to

Table 7.11 Water use variation by hydrographic basin

Hydrographic regions	Water availability	Water use 2005 (Million m^3)	Water use 2025 (Million m^3)	Growth (%)
Tocantins-Araguaya	624,106	2447	4826	97
Western Northeast Atlantic	187,327	363	705	94
The Parnaíba River	61,503	384	694	81
Eastern Northeast Atlantic	47,923	7853	8940	14
São Francisco River	96,557	4119	5946	44
East Atlantic	43,520	1830	2488	36
Southeast Atlantic	186,839	14,030	16,071	15
The Uruguay River	188,201	2910	3752	29

Scenario 1. 2005 and 2025

2025 is due to the irrigated land expansion. Pernambuco, Rio Grande do Norte and Paraíba would follow, but with a smaller share.

Our database allows tracking water supply by hydrographic basins. The simulation shows that the higher water demand increase would appear in the Tocantins-Araguaia hydrographic region, a 97% increase from the base year (Table 7.11). This hydrographic region also supplies part of the Mato Grosso state (basin 673) and the Goiás state (basins 664, 674 and 670), both located in Center-west Brazil.

The Eastern Northeast Atlantic hydrographic region is the one offering the least potential for expansion in water use among those that supply the Northeast region. Results show that this hydrographic region, which supplies Alagoas (basin 731), Ceará (basin 735), Paraíba (basin 733), Pernambuco (basin 732), Piauí (basin 736) and Rio Grande do Norte (basin 734), is the one that requires greater attention in irrigation expansions.

Finally, we stress the results on water demand increase in the São Francisco River basin. This is one of the most important hydrographic regions in Northeast Brazil, supplying the states of Bahia, Alagoas, Pernambuco and Sergipe. This river has its origins in Minas Gerais state, in Southeast Brazil, and flows across the Atlantic Forest, Cerrado and Caatinga biomes. Currently part of a national integration project aiming to divert part of the river flow to ensure water security for the semi-arid region of Northeast Brazil, this river is presently under severe pressure, mostly due to anthropization and poor territorial management. Our results show that the PNRH targets would put extra burden on this already fragile water basin.

7.5 Final Remarks

As in most nations, agriculture is the larger water user in the Brazilian economy. Irrigation expansion, than, has the potential to strongly affect water demand profile in the country. This is particularly true for irrigated sugar cane expansion, which, with extra 549,343 ha would raise water demand in 2949 million m^3 until 2025, seconded only by rice, with a 1808 million m^3 of water.

The results also point to the importance of analyzing water demand from a basin, instead of an aggregate, perspective. Actually, the climate and hydrogeological regional heterogeneity, including exploitation and activity risks, requires micro-regional oriented research and policies. The Northeast region in Brazil would be the region with the smallest expansion of irrigated area in three simulated scenarios. However, the impact over water resources in this region is noticeable, especially in Alagoas and Pernambuco states. Water demand required contemplating the whole cycle of irrigated crops in Northeast Brazil and increasing demand coming from population growth may not be consistent with the future water supply. For the scenarios and the time range evaluated in this study, however, the general picture do not allow to foresee generalized water supply problems, even in the northeast region.

Finally, note that our analysis does not include any climate change effect. The climate change scenarios for agriculture in Brazil point to a strong reduction in water availability in both the northeast and the center-west regions of Brazil. This could considerably change the reported results, both from the supply and demand sides. Demand should increase as an adaptation strategy, in face of a reduced rainfall that would reduce water supply in the water basins servicing the regions. This highlights the importance of expanding the integrated framework for water supply analysis presented in this paper to other basins in the country, a relevant research effort in Brazil.

References

Amorim Neto MS (1989) Balanço hídrico segundo Thornthwaite & Mather. Embrapa-Cpatsa, Petrolina

ANA (Agência Nacional das Águas) (2007) Disponibilidade e demandas de recursos hídricos no Brasil. Caderno de recursos hídricos 2:126. Brasília, Available at: http://biblioteca.ana.gov.br/index.html. Acessed in 2016

CPT (Comissão Pastoral da Terra) (2016) Conflitos no uso da água. Available at: https://www.cptnacional.org.br/biblioteca-virtual. Accessed in 2016

Doorenbos J, Kassam AH (1994) Efeito da água no rendimento das culturas [Effect of water on crop yield]. Campina Grande, UFPB, Paraiba, Brazil
dos Santos CV (2006) Política tributária, nível de atividade econômica e bem-estar: lições de um modelo de equilíbrio geral inter-regional [Tax policy, level of economic activity and welfare: lessons from an interregional general equilibrium model]. Ph.D. thesis, Escola Superior de Agricultura, Universidade de São Paulo, Piracicaba
Fachinello AL (2008) Avaliação do impacto econômico de possíveis surtos da gripe aviária no Brasil: uma análise de equilíbrio geral computável [Evaluation of the economic impact of possible outbreaks of avian influenza in Brazil: a computable general equilibrium analysis]. Tese (Doutorado em Ciências)—Escola Superior de Agricultura "Luiz de Queiroz", Universidade de são Paulo, Piracicaba
Fachinello AL, Ferreira Filho JBS (2010) Gripe Aviária no Brasil: uma Análise Econômica de Equilíbrio Geral [Avian Flu in Brazil: an economic analysis of CGE]. Revista de Economia e Sociologia Rural 48:539–566
Ferrarini ASF (2017) Avaliação setorial do uso da água no Brasil: uma análise de equilíbrio geral computável [Sectoral assessment of water use in Brazil: a computable general equilibrium analysis]. Tese (Doutorado). Escola Superior de Agricultura, Sao Paulo, Brazil
Ferreira Filho JBS, Horridge M (2014) Ethanol expansion and indirect land use change in Brazil. Land Use Policy 36:595–604
Horridge (2012) The TERM model and its database. In: Wittwer G (ed) Economic modeling of water: the Australian CGE experience. Springer, Dordrecht, pp 13–36
IBGE (Instituto Brasileiro de Geografia e Estatística) (2009) Censo Agropecuário 2006
Martins ESPR, Magalhães (2015) A seca de 2012–2015 no Nordeste e seus impactos [The drought of 2012–2015 in the Northeast and its impacts]. Parceiros Estratégicos 20:107–128
MI (Ministério da Integração e Instituto Interamericano de Cooperação para a Agricultura) (2014) Análise Territorial para o Desenvolvimento da Agricultura Irrigada no Brasil [Territorial analysis for the development of irrigated agriculture in Brazil]. Ministério da Integração e Instituto Interamericano de Cooperação para a Agricultura, Piracicaba, Brazil
MMA (Ministério do Meio Ambiente) (2006a) Caderno Setorial de Recursos Hídricos [Sectorial notebook of water resources]. Agropecuária. Ministério do Meio Ambiente, Secretaria de recursos hídricos, Brasília
MMA (Ministério do Meio Ambiente) (2006b) Plano Nacional de Recursos Hídricos: Síntese Executiva [National water resources plan: executive summary]. Secretaria de Recursos Hídricos, Brasília
MMA (Ministério do Meio Ambiente) (2006c) Plano Nacional de Recursos Hídricos: Águas para o Futuro [National water resources plan: water for the future]. Secretaria de Recursos Hídricos, Brasília
MMA (Ministério do Meio Ambiente) (2011) Desenvolvimento de Matriz de coeficientes Técnicos para Recursos Hídricos no Brasil [Development of a matrix of technical coefficients for water resources in Brazil]. Ministério do Meio Ambiente, Brasília
Pereira A (2005) Simplificando o balanço hídrico de Thornthwaite-Mather [Simplifying the T-M water balance]. Bragantia 64:311–313
Thornthwaite CW, Mather JR (1955) The water balance. Publications in climatology. Drexel Institute of Technology, New Jersey
Varejão-Silva MA (2006) Meteorologia e Climatologia. Versão Digital

Dr. Angel dos Santos Fachinelli Ferrarini has a Ph.D. in Applied Economics at Escola Superior de Agricultura "Luiz de Queiroz", University of São Paulo (USP), Brazil. She is currently a Professor at State University of Maringá (UEM). Her research focus is on the analysis of regional productive structure's development and regional spillover effects, including water use in Brazil, the policy scenarios of the National Water Resources Plan and their impact on water availability.

Prof. Joaquim Bento de Souza Ferreira Filho is a Senior Professor at Escola Superior de Agricultura "Luiz de Queiroz", University of São Paulo, Brazil, in the Applied Economics Graduate Program. His research focus mainly in the development of computable general equilibrium models for policy analysis in Brazil. In the last years, his work has been concentrated in the field of natural resources and socioeconomic consequences of environmental policies. He's been publishing in issues covering topics as deforestation, land use change, GHG emissions, water use, always with an inter-regional detail. He is also part of the Brazilian Panel on Climate Change, and a former president of the Brazilian Association of Agricultural Economics—SOBER.

Since 2002, when he finished his post-doctoral stage at the Centre of Policy Studies—COPS, Joaquim has been developing the TERM-BR model of Brazil, with several academic as well as consulting applications. It's an inter-regional, bottom-up model of the Brazilian economy that has been used in the assessment of many economic and environmental policies, like the Brazilian Intended National Determined Contribution to COP21, the Brazilian Forest Code, the Nagoya Protocol, the National Energy Plan, and the National Water Resource Plan.

Dr. Santiago Vianna Cuadra is a Research Fellow at the Brazilian Agricultural Research Corporation, EMBRAPA, Brazil. His research focus mainly in the representation of agriculture and natural eco-systems into biophysical models, and thereby explore the impacts of land use changes on climate and water cycle as well as the impacts of climate change on crop yield. Present research activities includes development and application of agro-ecosystem models, development new food and bioenergy crop models, coupling agro-ecosystem to climate models, and assessments of climate impacts over bioenergy and food security.

Dr. Daniel de Castro Victoria is a Researcher at the Brazilian Agricultural Research Corporation, EMBRAPA, Brazil, has an undergraduate degree in Agronomic Engineering from University of São Paulo (2002), a masters degree from University of São Paulo (2004) and doctors degree from Center of Nuclear Energy in Agriculture (CENA—USP—2010). His research topics include geoprocessing, remote sensing, agro-meteorology, land use and cover change and hydrological modeling. His works include large-scale agriculture mapping and hydrological modeling of the amazon basin.

Prof. Mark Horridge is an expert on CGE databases and software. He has helped to develop CGE databases for Albania, Australia, Brazil, China, Finland, Indonesia, Jersey, Kazakhstan, New Caledonia, New Zealand, Philippines, South Africa, Taiwan, Thailand and Vietnam. He is the Director of GEMPACK software, a specialist CGE package used by over 400 organisations in over 70 countries. He is primarily responsible for developing a number of GEMPACK tools. He is the creator of the TERM model and its predecessor MMRF, and also contributed to the development of the dynamic MONASH model. He has developed a unified database that underlies these Australian models. Professor Horridge has developed much teaching material for short CGE courses, and is in high demand as an instructor as these courses around the world. For many years he has also instructed at the GTAP short courses. In 2007, he became an inductee into the GTAP Hall of Fame for his development of new software and tools for implementation and analysis of GTAP data and simulations.

Chapter 8
Distributing Water Between Competing Users in the Netherlands

Jason F. L. Koopman, Onno Kuik, Richard S. J. Tol, Marnix P. van der Vat, Joachim C. Hunink and Roy Brouwer

Abstract The Netherlands is a delta country where water is usually abundant. Large investments in water infrastructure aim to prevent flooding, maintain shipping transport routes, irrigate farmland and ensure the health of polder lands and nature. During the limited periods when water is scarce, agriculture is low on the priority list for water allocation: farmers may be restricted in expanding irrigation operations or be even temporarily forbidden from using the equipment already installed. This comes at a cost to agricultural production. Water in this context is a unique economic input that is not privately owned, not always scarce, and not always allocated according to market principles. Nonetheless, the framework of a computable general equilibrium model (CGE) can be very effective in assessing economy-wide changes from periods of water scarcity and weighing this against policy initiatives to reduce water scarcity. In this chapter we explore adaptation possibilities to water scarcity from climate change with a particular focus on the challenges of interpretation of the CGE methodology for water in the context of the Netherlands.

This chapter reproduces with permission substantial portions of Koopman et al. (2015, 2017).

J. F. L. Koopman (✉)
Wageningen Economic Research, The Hague, The Netherlands
e-mail: jason.levin-koopman@wur.nl

O. Kuik · R. S. J. Tol · R. Brouwer
Department of Environmental Economics, Institute for Environmental Studies, Vrije Universiteit Amsterdam, Amsterdam, The Netherlands

R. S. J. Tol
Department of Economics, University of Sussex, Falmer, UK

R. S. J. Tol
Department of Spatial Economics, Vrije Universiteit Amsterdam, Amsterdam, The Netherlands

M. P. van der Vat · J. C. Hunink
Deltares, Delft, The Netherlands

R. Brouwer
Department of Economics and The Water Institute, University of Waterloo, Waterloo, Canada

G. Wittwer (ed.), *Economy-Wide Modeling of Water at Regional and Global Scales*, Advances in Applied General Equilibrium Modeling,
https://doi.org/10.1007/978-981-13-6101-2_8

Keywords Water scarcity · Climate change · Computable general equilibrium · Water markets

8.1 Introduction

One of the climate change scenarios that has been developed for the Netherlands predicts hotter and drier summers and a substantial drop in river discharge in summer. This is expected to lead to increased water scarcity with potentially detrimental economic and environmental effects. The summer of 2018 was one of the driest years on record and according to a recent review by Coumou et al. (2018) this hot and dry extreme weather could occur more often as part of a persistent pattern in the mid-latitude region of the globe. Global climate change exerts significant pressure on the way that we allocate our limited water resources across different water uses and user groups. Different countries apply different allocation rules, varying from national top-down command and control policies to local or regional water markets and transboundary river basin treaties. In examining adaptation responses to climate change, such as updating water infrastructure, economic analysis can play an important role in reducing costs and improving efficiency (Hughes et al. 2010).

Traditionally questions of water distribution for economic use have been the realm of engineering models. More recently hydro-economic models which include economic principles, in the form of water demand functions, have been used to optimize the allocation of existing water resources (Harou et al. 2009). In the past two decades, Computable General Equilibrium (CGE) models have seen increasingly wider use in the economic assessment of water allocation and policy Brouwer and Hofkes (2008). The use of CGE models to examine water policy dates back at least to Berck et al. (1990) who examined the economic costs of transferring water from agricultural to recreational use in California. Later papers have added increasing levels of water heterogeneity in time (Goodman 2000), in space (Diao et al. 2005; Liu et al. 2016) as well as distinguishing between surface and groundwater (Diao et al. 2008) and desalinated water (Gomez et al. 2004).

Ponce et al. (2012) reviewed this literature, and among their findings is the general lack of detail in non-agricultural sectors and in industrial water-using sectors in particular. Further, they found that most studies essentially examine a loss of water productivity instead of an explicit loss of water availability. Some notable exceptions to this are Gomez et al. (2004) and Goodman (2000), who examine market water allocation in the Balearic Islands and the Colorado river respectively. In these studies there is a "raw water" endowment which is transformed into agricultural or municipal water before use in these sectors. Goodman (2000) also makes the distinction between water available at various times of the year and allows the market price of water to fluctuate with the demands in time of the various users.

Adaptation to climate change is defined by the IPCC (2014) as "the process of adjustment to actual or expected climate and its effects". Several types of adaptation can be distinguished. A well-known distinction is that between planned (anticipatory)

Table 8.1 The three studies examined in this chapter, the adaptation mechanisms and the model characteristics central to the study

The study and model	Adaptation mechanism	Water modelling characteristics	Assumptions on water use
1st study uses model 1 (GTAP-W)	Autonomous market adaptation	Irrigation water as explicit endowment in agriculture	Explicit water use restricted to agriculture. Additional abstraction of groundwater not allowed
2nd study, uses model 2 (distinction between surface and groundwater)	Increased investment in irrigation infrastructure	Surface and ground water as separate endowments for irrigation in agriculture	Explicit water use restricted to agriculture. Groundwater abstraction limited only by irrigation infrastructure
3rd study, uses model 3 (physical water markets)	Explicit water market across multiple sectors	Water as explicit endowment in agriculture, industry and public water services. Water market allows for trading between sectors	Explicit water use in agriculture, industry and public water services. Additional abstraction of groundwater not allowed

and autonomous (reactive) adaptation (Smit et al. 2000). In this chapter we explore market adaptation responses to water scarcity from climate change in the Netherlands by examining aspects of three studies, each of which uses a CGE model. The first (Koopman et al. 2015) explores the extent of the autonomous market response. That is autonomous adaptation by economic agents that is triggered by market/price signals. The second study (Koopman et al. 2018, in preparation), explores the planned adaptation of increased investment in irrigation water infrastructure in the agricultural sector. The third study (Koopman et al. 2017) explores the planned adaptation of introducing explicit water markets between industry, agriculture and public water services (Table 8.1).

In the first and third studies, we focus on surface water that is supplied by rivers and through precipitation. We assume that groundwater deposits are not (further) depleted i.e. that renewable groundwater is used sustainably, so the groundwater level plays no role in the analysis. In the second study we relax this assumption and assume that groundwater abstraction is limited only by irrigation infrastructure rather than physical water availability. The reality will most likely be somewhere in the middle of these two extremes as it is expected that groundwater abstraction can be on average increased slightly but not without limitation. This assumption on the availability of ground water in the second study leads to a positive impact on overall crop growth in the Netherlands from a warmer climate, while in the first and third studies the direct impact of climate change on crop growth is negative. In all three studies there is a separate public water services sector that supplies drinking water to all sectors and households. The raw water needed to supply this sector is considered in the third study, and left out of the analysis in the first and second studies.

8.2 Water Use in the Netherlands and the Impacts of Climate Change

The Netherlands is a delta region and water is deeply interwoven with the Dutch economy and way of life. Two major rivers flow into the Netherlands. The Rhine from Germany and Switzerland and the Meuse from Belgium and France. These rivers provide many economic services which range from allowing commercial and recreational shipping and groundwater recharge, to providing irrigation water and preventing the sea water from infiltrating the groundwater which would otherwise degrade the productivity of farmland. The countries that share these two rivers are linked economically and politically, and they draw much of their water from the same sources.

In the Netherlands, agriculture accounts for a relatively small 3% of total water withdrawals (Eurostat 2014; Hoogewoud et al. 2013), compared with 24% of total water withdrawals in Europe (Collins et al. 2009) and 80% globally (Rosegrant et al. 2002). However, very little abstracted water is actually consumed in other uses, roughly 10% for various manufacturing industries and less than 2% for cooling in electrical power generation (Scharf et al. 2002), whereas agriculture consumes approximately 70% of abstracted water (Collins et al. 2009).

Even though agriculture is a relatively small water user in the Netherlands; it is the focus of this chapter. The reason for this is that agriculture is considered a low priority economic sector as opposed to the other economic uses of water and in times of water scarcity other sectors will have priority for the available water. Hence any scarcity will be first felt by the agricultural sector. This prioritization is in keeping with the "verdringingsreeks", a priority use list for water uses and users in the Netherlands (Ministerie van Verkeer en Waterstaat 2009).

Water users in the Netherlands can draw their water from both groundwater and surface water sources. Groundwater and surface water sources each have their own set of hydrological characteristics and alternative uses that are taken into account when deciding what restrictions should be placed on abstraction for irrigation. For example, the supply of surface water reacts much more immediately to changes in precipitation, while groundwater reacts more slowly and can be used as a short term buffer in times of relative drought. Moreover, the health and stability of the soil in many areas of the Netherlands is critically dependent on the level of groundwater as are many important nature areas and maintaining current levels of groundwater is therefore one of the main priorities of water managers (Ministerie van Verkeer en Waterstaat 2009).

For a low-lying country such as the Netherlands, sea and river flooding is generally considered to be the most serious threat from climate change. Yet, two of the four climate change scenarios that were drafted by the Netherlands Royal Meteorological Institute (KNMI) predicts substantially warmer and dryer summers and a significant decrease in summer discharge by 2050 of two of the Netherlands' major rivers, the Rhine and the Meuse (van den Hurk et al. 2006; te Linde 2007). The threat of water scarcity brings challenges to many aspects of Dutch society, from the supply of drinking water to production in agriculture and industry, the health of ecological systems, and the stability of water infrastructure such as levees and dikes (Klijn et al. 2012).

Climate change may have mixed impacts on global markets. For example, countries around the Mediterranean might have a serious reduction in agricultural production from higher temperatures and reduced rainfall. In contrast, warmer temperatures might create favorable agricultural conditions in Greenland and other northern areas. Similarly, rising sea levels might threaten many world ports, but warmer temperatures might also open up new trade routes in the arctic, reducing transportation costs. Therefore for simplicity, in this chapter we restrict our attention to the impact of climate change on water scarcity to the Netherlands. Without knowing the structural changes in global trade patterns from climate change we assume that the pattern of global trade, i.e. the demand for Dutch exports and the availability of imports into the Netherlands, will remain unchanged by the changing climate.

8.3 Models and Data

8.3.1 Models

The models used in the three studies are extensions of the GTAP-W model first introduced by Calzadilla et al. (2010). The GTAP-W model from Calzadilla and co-authors is itself an extension of the GTAP model (Hertel 1997).

GTAP-W extends the GTAP model by including more detail into the land endowment for agricultural producers (which are the exclusive users of the land endowment), splitting the original endowment into rain fed, irrigated, and pasture land based on the value of production of these commodities and then further splitting off irrigation water from the irrigated land endowment. All of the new land and water endowments in GTAP-W inherit the partial mobility parameter from the original land endowment in GTAP. In GTAP-W, in the lowest nest of the CES production function, crop farmers determine the level of irrigation based on the relative prices of land and irrigation water and the technical ease of varying the level of irrigation water on a given piece of land. In higher nests of the production function, the irrigated land is combined with capital, labor, and intermediate goods (seeds, fertilizers, pesticides) to produce an output like wheat or sugar beets. The CES production nest for the GTAP-W model is shown in Fig. 8.1 focusing on aspects most relevant to this paper.

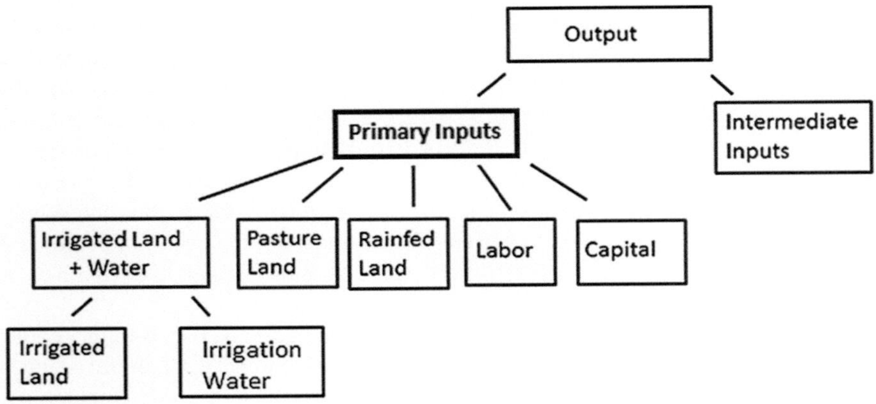

Fig. 8.1 The CES production nest for agriculture in GTAP-W (model 1). *Source* Koopman et al. (2015), Fig. 1

For the second study we use an extension of the GTAP-W model. Here the irrigation water and irrigable land endowments from GTAP-W have been replaced by four endowments: irrigation from groundwater (GW), land irrigated by groundwater, irrigation from surface water (SW), and land irrigated by surface water. Further, the structure of the CES function has been adjusted such that all land endowments related to crop agriculture are placed in a single nest "crop land" at the same level as labor, capital and pasture land. The crop land aggregate is then further split into rain fed land and the irrigated land composite. The irrigated land composite is further split into a surface water composite and a ground water composite, which are then each split into an irrigable land and irrigation water component respectively. The new production structure for model 2 is shown in Fig. 8.2.

In model 3, we use a separate extension of the GTAP-W model. We separate the value of physical water from the rest of the irrigation water endowment. The irrigation water endowment from the GTAP-W model has been replaced by an irrigation water composite which is the combination of physical water volumes and dedicated irrigation water capital. Figure 8.3 shows the decomposition of the irrigation water composite in model 3. The rest of the CES production nest for agriculture including the irrigated land endowment and the irrigated Land + water composite are the same as in model 1 (see Fig. 8.1).

In contrast to agriculture, the manufacturing and public water services sectors combine water with capital (not land). The production for these sectors is shown in Fig. 8.4. The possibility of substitution between water and capital is in line with the findings of Renzetti (1992) and Dupont and Renzetti (2001) who assert that intake water may be a substitute for recirculation water which is more capital intensive. Alternative forms of irrigation—flood, sprinkler, drip—also have different capital and water intensities. The CES for the crop sectors however has capital as an direct input in the primary inputs nest and not directly as a substitute for water. The value

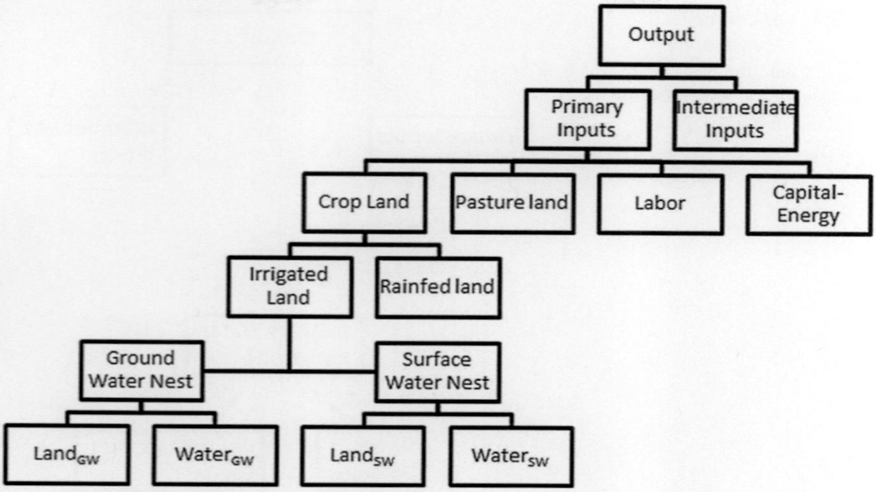

Fig. 8.2 The production structure used in model 2

Fig. 8.3 The decomposition of the irrigation water composite in model 3

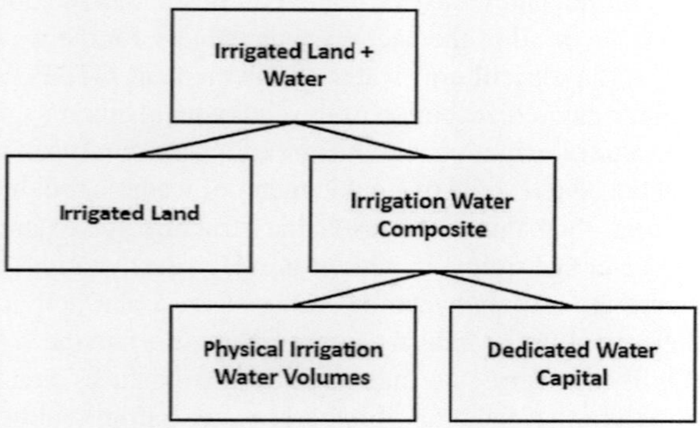

of the industry water composite is then defined as the firm expenditure on water abstraction, purification prior to use as well as discharge after use. The activities represented by the industry water composite can be substituted for expenditure on recirculation water which is assumed to be part of the capital endowment. These assumptions are in line with Goodman (2000) and Gomez et al. (2004) who combine water with capital in their models in a similar way.

We further extend model 3 by accounting for volume flows of water between sectors. This addition is necessary to insure a physical water balance when water is exchanged between very different types of use. In GTAP-W as presented in Calzadilla et al. (2010), the agricultural water endowment (renamed here the agricultural water composite) represents all of the benefits (and also the expenditure) of irrigation for production. This includes everything involved in irrigation, not only physical water but also irrigation equipment. We make the same assumption that the values of the

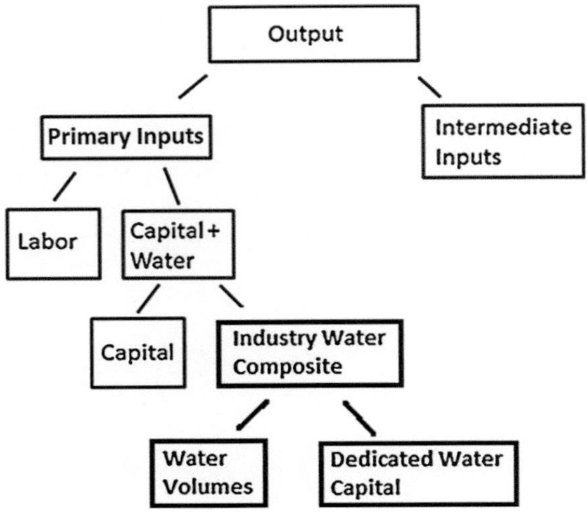

Fig. 8.4 The constant elasticity of substitution (CES) production nest for non-agricultural activities in model 3. *Source* Koopman et al. (2017), Fig. 2

agricultural and industrial water composite include not only the water itself but also the value of all of the necessary machinery for the water activity.

As the agricultural water endowment in GTAP-W is measured in millions of dollars, any redistribution of the endowment among agricultural sectors redistributes the value of irrigation without an explicit accounting of water volumes. If one assumes that the added value of a cubic meter of water is roughly the same for all agricultural sectors, then this is a reasonable structure for examining a market for water (or for water use rights) in agriculture. However, study 3 examines water redistribution where the ratio between the value of the water activity in production (the value of the agriculture or industry water composite) and the volume of water involved in that activity can vary quite substantially between uses, and this needs to be accounted for to ensure that the total volume of water before reallocation is the same as the total volume after reallocation.

To separate the value of physical water from the rest of the agricultural water endowment, we have changed the name of irrigation water 'endowment' to irrigation water 'composite', and split it up further into physical water volumes and dedicated agricultural water capital (see Fig. 8.3). Similarly, for the public water services sector and the manufacturing sectors, we first split off the industry water composite from the rest of the capital endowment where the industry water composite includes the value of all expenditures on water-related activities of abstraction, purification, use in production, and discharge. We then further split the industry water composite into the value of physical water volumes and dedicated industry water capital, which represents the value of water equipment (see Fig. 8.4). Dedicated water capital is immobile. Physical water volumes are mobile in principle, but the mobility is restricted in various policy scenarios to simulate the various water market alternatives. The physical water volume endowment is combined with dedicated water capital with Leontief production technology (no substitution allowed). The immo-

Table 8.2 Sector aggregation used in the reporting of results of the three studies

Sector aggregation
Wheat
Cereal crops
Vegetables and fruits
Sugar beets
Other crops
Public water services
Manufacturing
Services and other

bility of dedicated water capital and the Leontief production technology between dedicated water capital and physical water volumes means that the value of the water composite is completely determined by the amount of physical water available and that a percentage reduction in the amount of available water for production results in the same percentage reduction in the water composite.

In the Netherlands, there is no market for self-abstracted raw water. Determining the exact value share of the water composite that should be attributed to the water volume endowment if the firms were to be granted property rights over the water that they use was beyond the scope of the study. See Appendix 2 of Koopman et al. (2017) for details on the procedure used to estimate the relative value shares of the endowments of physical water volumes.

8.3.2 Data

All three studies use version 6 of the GTAP database (Dimaranan 2006). This database has 87 regions each with 57 sectors, which together describe the global economy in 2001. The sector aggregation examined in this chapter is shown in Table 8.2.

The methodology for splitting the value of the land endowment from the GTAP data follows Calzadilla et al. (2011a, b), where pasture land is the value of the original land endowment used by the Animals sector, the value of rain fed land is split from the value of the irrigated land–water composite by the fraction of rain fed crop production to total production. The value of irrigation water is split off of the land–water composite using the ratio of rain fed yields to irrigated yields as shown in Sect. 3.4. Irrigation water is a composite for everything involved in creating additional yields in irrigated land. The value attributed to irrigation water (the irrigation water composite in model 3) defined in this way is not simply the water itself, but everything involved in irrigation, including equipment.

8.3.3 Elasticities of Substitution

The elasticity of substitution between the irrigated land endowment and the agricultural water composite is 0.1. The elasticity of substitution between capital and the industrial water composite for the manufacturing sectors is 0.5 and 0.1 for public water services. The elasticities of substitution were calibrated such that a 10% reduction in water availability would induce the same price-water elasticities that are given in Rosegrant et al. (2002).

In model 2, we assumed that irrigated and rain fed agriculture produce the same (or very similar) crops and therefore an increase in irrigated agriculture would not necessitate a similar increase in the production of the rain fed product. This would not be the case for example if within the aggregated vegetables and fruits sector irrigated agriculture was primarily apples and rain fed agriculture was primarily carrots. Therefore, it was assumed that irrigated land and rain fed land are nearly perfect substitutes and consumers were indifferent between the two. The same assumption was made for the groundwater and surface water aggregates. Therefore the elements of the Crop land nest and in the irrigated land nest have a very high substitution elasticity of 3. The elements of the groundwater nest and the surface water nest were calibrated such that a 1% increase in the productivity of irrigation water (surface or ground) would result in a higher and not lower demand for the irrigation water endowment. This resulted in each having an elasticity of 1. The elasticities of substitution among the elements of the primary inputs nest differ per sector and were taken from the standard GTAP model. It was assumed that it is not possible to substitute between intermediate and primary inputs.

8.3.4 Ground and Surface Water

In the second study, the value of the rain fed land and the irrigated land composite, as well as the value of the surface water and ground water nests (see Fig. 8.2) are all derived from the value share of production using the same methodology as Calzadilla et al. (2011a, b).

For both the groundwater and surface water nests, irrigated land is split from irrigation water by comparing the ratio of irrigated yields to rain fed yields, where the value of the additional yields from irrigated production is attributed to water. The equation for splitting the value of irrigation water from irrigated land for each crop in each region is shown in Eqs. 1 and 2 below (modified from Calzadilla et al. 2011a). For more detail on GTAP-W see Calzadilla et al. (2011a, b).

$$Lnd_{j,r} = LndWatC_{j,r} * \left(\frac{YRF_{j,r}}{YIRG_{j,r}} \right) \tag{8.1}$$

$$Wat_{j,r} = LndWatC_{j,r} * \left(1 - \frac{YRF_{j,r}}{YIRG_{j,r}} \right) \tag{8.2}$$

where:

LndWatCj,r	The value of the Land-Water composite
Watj,r	The value of the Water endowment
Lndj,r	The value of the Irrigated Land endowment
YRFj,r	The rain fed yield
YIRj,j	The irrigated yield

for all crops j and for all regions r.

Therefore the value of land irrigated by groundwater is split from the value of groundwater irrigation by comparing the yield ratio between groundwater irrigated agriculture to rain fed agriculture, and applying Eqs. 1 and 2. The surface water aggregate is split in the same way.

For the value of the production of rain fed agriculture, and irrigated agriculture by ground water and surface water in the Netherlands we used data from the Netherlands Hydrological Instrument (NHI) (Hoogewoud et al. 2013) and the AGRICOM crop model (Mulder and Veldhuizen 2017). The AGRICOM model provided data on how yields would change in the Warm+ climate scenario described in van den Hurk et al. (2006). The NHI model includes detailed information on river flows as well as the needs of nature and other demands on water use. It therefore assumes restrictions on the surface water available for irrigation. There are no policy restrictions on ground water abstraction, and the NHI model assumes that the only restriction on the volume of abstracted ground water is the capacity of the current irrigation infrastructure. In contrast, the Deltares Delta Program Phase 3 data, which was used in the first and third study, are based on calculations that do not allow an increase of groundwater abstraction. The more conservative assumptions regarding ground water abstraction in the Delta Program Phase 3 resulted in yield losses in the warmer climates studied, while the assumed ground water availability in the NHI data used in the second study combined with an increase in irrigation water abstraction allows for an increase in yields in the warmer climate of the Warm+ scenario.

8.3.5 Splitting Water and Capital

In order to determine the value of the water composite for the manufacturing sectors in the third study, we used a survey by Scharf et al. (2002) of Canadian manufacturers, which details expenditure on water extraction, treatment, recirculation, and discharge for several manufacturing sectors. We transferred these expenditures to our model regions by using the number of employees per manufacturing sector from Scharf et al. (2002) and Eurostat (2012) as a scaling factor. The value of the industry water composite for the non-European region was determined by imposing the same ratio of the value of the water composite to the value of capital as the rest of Europe region. In the Manufacturing sectors in the Netherlands, the share of expenditure on water abstraction, treatment and discharge compared with the total expenditure of the sector ranges from the low end of 0.05 and 0.1% in other manufactures and food products respectively to a high of 0.5% in metal manufacturing. For the value of the water composite in the public water services sector, we relied on Teeples and Glyer (1987) who estimated a constant cost share of raw water of 18%.

For volumes of water used in the manufacturing and public water services sectors in the third study, we used data from the Eurostat website (Eurostat 2014). For the Dutch irrigated water volumes we used data from Hoogewoud et al. (2013). These differ substantially from the Dutch irrigation data from the Eurostat website, but we used Hoogewoud et al. because it is consistent with the Delta Program Phase 3 data which we used for the value of Dutch crop production and the direct effects of climate change on agriculture.

In estimating the water volumes used by industry, we ignore water used for cooling, which takes place mostly in the energy sector, but also in certain types of manufacturing. Water used for cooling is a process which involves abstracting large volumes of surface water, but returning virtually the same quantity and quality, only slightly warmer. As water for cooling involves very little consumptive use, it does not necessarily involve a trade-off between users. An exception is if the water temperature is already quite high, then warmed water could affect the ecosystem where it is discharged. Trade-offs between water for environmental and economic use and the resultant feedbacks of water quality on economic use are beyond the scope of this paper (see Brouwer et al. 2008; Dellink et al. 2011; Zhu and van Ierland 2012 for studies that include water quality in assessments of water for economic use).

8.3.6 The Nature of the Water Endowment

The water endowment represents the value of the sector's abstraction of raw water and equipment required for abstraction, use and discharge. The value represented by the water endowment does not include payments to the public water services sector for drinking water.

The Netherlands is a delta country where water is usually plentiful; it has relatively high or low levels of irrigation depending on how it's defined. Dutch farming is capital-intensive and there is a relatively high amount of irrigation equipment but they are only activated on unusually hot and dry days, and there is less water used in farming compared with other countries. Therefore even though there is a large proportion of farms equipped for irrigation, relatively little water is used in irrigation.

The Netherlands has a heavily developed infrastructure system for controlling and directing river water. It also has significant shallow groundwater resources. Most days irrigation water will not be restricted and there is excess supply but on certain hot dry days of the summer the demand for water will exceed the pump capacity or capacity of the canal network to deliver the water to where it is needed. There may also be an irrigation reduction or ban imposed on farmers by the water authorities to meet other water demands. Such demands include regulating water levels in the polders, ensuring sufficiently high groundwater levels for nature and preventing salt water intrusion into the fresh water systems. In these cases, water restrictions due to policy or physical infrastructure results in a water demand exceeding supply. This is leads to a reduction in potential growth for crops.

The water endowment in this study is defined as the value of the ability to abstract and use water when needed, under the current infrastructure. For agriculture the value of the endowment is reflected in the value of the additional yield (see Eqs. 1 and 2).

In the Netherlands as in most places the use of raw water is not charged volumetrically. The activities of the water managers are paid for by local taxes. Large scale abstraction is subject to a license, but there is no tariff per unit of abstraction and therefore there is no water price as such. There is a shadow price of water however, which is the value added that would be created by increasing the supply of water by one unit for a particular activity. If water is abundant then this shadow price is negligible. Still if at a particular time and place, freshwater becomes scarce (meaning that not all demand can be satisfied at present conditions), the shadow price of water will rise.

As freshwater has a shadow value in times of scarcity it is a natural resource that provides economic as well as environmental services (Briscoe 2005). Young and Haveman (1985) already noted more than 30 years ago that water has unique physical properties, complex economic characteristics, and important cultural features that distinguish it from other resources. The idea that water resources management can benefit from economic principles can also be defended (Briscoe 2005), but should always take account of other, non-economic values that may restrict the scope of these principles.

As there is no market for physical water in the Netherlands we do not have data on a water price or even quantities of water used in place and time. Therefore instead of conceiving of an explicit water endowment as the value of physical water for a specific sector, we define the water endowment loosely as the value of the water activity for the sector similar to Calzadilla et al. (2010). In the case of agriculture, the value of the water endowment is calculated as the difference in yield between rain fed and irrigated land crops. Therefore, the endowment irrigation water is a composite for everything involved in creating additional yields compared with its rain fed counterpart. The value attributed to irrigation water defined in this way is not simply the water itself but everything involved in irrigation, including equipment and labor.

By defining the value of the agricultural water endowment as the added benefit to the sector of irrigation, we acknowledge that this estimate is an upper bound on the actual sector expenditure on the water activity which may be much smaller. This is most likely the case in the Netherlands as the water is not paid for and there is a lot of public infrastructure to ensure water reaches the farm. In this case the difference between expenditure on water activity and the added value from the water activity should be attributed to the value of irrigated land endowment, instead of the irrigation water endowment. So this procedure would underestimate the value of irrigated land and overestimate the value of irrigation water endowment from the perspective of firm expenditures. If a volumetric water price were implemented, then the firm might be willing to pay if the new cost of the irrigation including the volumetric price as well as equipment etc., did not exceed the current estimation of the water endowment. However the value of irrigated land would then be expected to fall. If the total expenditure on the water activity would exceed the benefits (determined by the difference between irrigated and rain fed yields), the farm would use less water either by optimizing the use of irrigation equipment to increase water efficiency or by irrigating less.

In contrast to the value of the water endowment for agriculture, the value of the water endowment for industry and public water services is split from the value of capital and defined by the expenditure on abstraction, purification prior to use and discharge. By defining the value of the water endowment in this way for the industrial sectors and public water services, this estimate is then a lower bound on the value of the water activity for the sector involved. That is, if a volumetric water price were implemented then the water using sectors might well be prepared to pay.

8.4 Selected Adaptation Scenarios

8.4.1 Water Scarcity Scenarios from Crop Models

The three studies examined in this chapter involve variations of the future dry climate based on the Warm+ climate scenario (van den Hurk et al. 2006). The first study approximates the direct impact on crop yields from the Warm+ climate scenario by implementing the direct crop yield response to the extremely dry climate of 1976 based on the Deltares Delta Program Phase 3 results, which assumes a maximum level of groundwater withdrawals and predicts a reduction in crop yields and land fertility in the Netherlands as a response to the dry climate. The second study approximates the impact on crop yields from the Warm+ climate scenario directly, using the NHI data (Hoogewoud et al. 2013) which assumes no restrictions on groundwater extraction resulting in a positive yield response to the increase of temperature under the Warm+ climate. This yield increase rests entirely on crops having enough water to take advantage of the warmer temperatures. In the Warm+ climate scenarios with particularly dry years using the NHI and AGRICOM models, crop yields will decrease since there is not enough irrigation water under current infrastructure to satisfy the needs of the plants. The third study implements the Warm+ climate scenario directly but uses the same groundwater assumptions and crop response model as the first study (Delta Program Phase 3), resulting again in a yield decrease for all crops.

Table 8.3 Climate impacts on the rain fed and irrigated land productivity in studies 1 and 3, shown in percent change

	Wheat		Cereal crops		Vegetables and fruits		Sugar beets		other crops	
Study	1	3	1	3	1	3	1	3	1	3
Irrigated land	Na	Na	Na	Na	−15	−4	−14	−12	−3	−10
Rain fed land	−13	−10	−13	−10	−37	−23	−27	−20	−27	−20

Na Not applicable
Source Based on Koopman et al. (2015, 2017), Tables 3 and 4 respectively

8.4.2 Changes to Endowments to Reflect Productivity Changes

The data sources used in the studies provide the impacts on crop production from the Warm+ climate. In the first and third studies these changes in irrigated crop production were translated to an impact on the agricultural water and land endowments using a method described in Koopman et al. (2015), by fixing prices for all elements of the CES crop production function with the exception of water and land to estimate the required change to irrigated land productivity which would result in the estimated reduction in crop production from the Delta Program Phase 3 data given a change in the quantity of the water endowment. The percent change production of the rain fed crop was directly implemented as the percent change in the productivity of rain fed land.

In the dry climate crops need more physical water to achieve the same yield. In the first and third studies the direct crop response to the dry climate was reduced crop yields and so we reduced the quantity of the water endowment by the percent difference between the volume of water required to achieve the pre-drought yields and the actual volume of water abstracted. In that dry climate the abstracted water was actually higher than the abstracted water in the pre-drought climate, however the water required to achieve the same yields was significantly higher still and therefore the quantity of the water endowment was reduced. This reduction of the water endowment was not enough on its own to reduce the crop output to the target levels in the calibration of the CES and so the productivity of irrigated land was reduced as well to achieve the remaining reduction of crop growth. Table 8.3 shows the changes to the irrigated and rain fed land productivity as a result of the warmer climate scenarios for the studies 1 and 3. The quantity of the irrigation water endowment was reduced by 22 and 11% in studies 1 and 3 respectively.

In the second study we chose a slightly different approach. Recognizing that the growth of the plants in the water scarce environment is dependent on water, we attributed all change in crop growth to the water endowment leaving the productivity of irrigated land unaffected. We then attributed the quantity of the irrigation water endowment to irrigation capital (canals, dams, water pumps and on farm irrigation systems) while the changes in climate and water availability were attributed

Table 8.4 Climate impact on the endowment productivity in the second study

	Climate impact for 2nd study, % change in productivity				
	Wheat	Cereal crops	Vegetables and fruits	Sugar beets	Other crops
Surface Water endowment	Na	Na	49	79	50
Ground water endowment	Na	Na	45	77	54
Rain fed land endowment	6.8	6.8	−4.7	−0.5	−4.3

Na Not applicable

to changes in the productivity of the water endowment and not the quantity of the water endowment. Table 8.4 shows the changes to the productivity of the surface and groundwater endowments and rain fed land as a result of the Warm+ scenario using the data from the NHI and AGRICOM models. The large changes in productivity of the irrigation water endowments reflect the fact that they represent a small portion of the total share of the surface (or ground) water nest (see Fig. 8.2), therefore even a small change in the irrigated crop production requires a large change in the productivity of the water endowment according to the calibration procedure as discussed in Koopman et al. (2015).

Here again the productivity of the water endowment is not necessarily the productivity of the physical water. In fact plants would need more water in the hot and dry Warm+ climate then they do in the current climate to produce the same yields. The productivity of the water endowment in the second study is defined as the productivity of the existing irrigation systems on the farm and the infrastructure (canals, dams) which brings water to the farm. Under the normal climate conditions a farmer might only turn on the irrigation system during the few dry periods and the rest of the time it is not used. Under the Warm+ climate conditions the same irrigation system will be used much more often, bringing more water to the plants, not only on a given day but also on more days total. This results in higher yields for the plants which we ascribe purely to the water endowment. In this way it is the productivity of the irrigation activity which is generating higher yields rather the productivity of physical water.

8.4.3 Selected Adaptation Responses to Water Scarcity

In the first study we compare the direct damage to agricultural production from the warm climate provided by the crop model of the Delta Program Phase 3 without economic feedbacks to the total economic impact of the same climate shock as calculated in model 1 which does allow for economic feedbacks. In this way the aim of study 1 is to assess the extent of autonomous adaptation to the climate shock in agriculture.

In the second study we use model 2 to compare the economic impact of the Warm+ climate scenario with autonomous adaptation to a scenario with additional investment in surface water irrigation infrastructure as planned adaptation. In this way the aim of study 2 is to assess the additional economic impact of irrigation investments as planned adaptation as opposed to only autonomous adaptation.

In the second study the crop response to the Warm+ climate scenario from the NHI and AGRICOM models suggest that if the crops which were irrigated by surface water were to receive the optimal amount of water for growth then the output would increase by over 14% compared to the Warm+ scenario crop yields under current irrigation infrastructure. In an additional scenario in the second study, using model 2 (see Fig. 8.2), we explore this possibility by increasing the quantity of the surface water endowment by 65%. In this case, the output of agriculture irrigated by surface water would increase 10% absent economic feedbacks and adjustments of other inputs by farmers. The surface water endowment was calibrated to achieve the 10% increase in production using the same assumptions on the CES production function used in studies 1 and 3 and described in Koopman et al. (2015).

In the third study, the reduction in the quantity of the water endowment as a result of the Warm+ climate was applied to the physical water volumes endowment (see Figs. 8.3 and 8.4) instead of the aggregated water composite as in studies 1 and 2. By distinguishing between water volumes and dedicated water capital in the third study we were able to ensure that in a water market between the two users that it would be physical water that would be traded and not value of water equipment. As we were unable to obtain data on the quantity changes in demand and supply of water for the manufacturing and public water services sectors we choose to reduce the water endowment of industrial water users by the same amount as the agricultural water users.

The Warm+ climate scenario predicts increased water availability in the winter. We assume, however, that extra winter water cannot be used in the summer in excess of what is already being stored under the current infrastructure. The results from the Delta Program Phase 3 suggest an 11% loss in effectively available irrigation water in the summer (we assume that all irrigation occurs in the summer). In study 3 we assume that the Warm+ scenario causes a proportional loss of water available for manufacturing and public water services, implying that there will be an 11% loss to the entire summer water supply available for economic use. We assume that the public water services, and manufacturing sectors have a constant requirement for water use throughout the year, and so, the 3-month summer water requirements of these sectors is 25% of their yearly water requirement, and the value of the summer water endowment is 25% of the value of the water endowment in 2001. To ensure that only summer production is affected by the summer water reductions, we divided all water-using sectors into a summer fraction and a rest of the year fraction (with the exception of the irrigated agricultural activities whose loss of yearly output was calibrated to a loss of summer irrigation water from Delta Program Phase 3 results.

Only the summer fraction of manufacturing and public water services is affected by water scarcity.

In the third study we compare 4 scenarios of water market allocation in the warm climate. All scenarios are implemented in model 3 and so include autonomous adaptation. The aim of study 3 is to assess the importance of water markets (or the allocation of water according to its shadow price) as planned adaption to climate change.

In these studies, the term water markets is defined rather loosely as a mechanism which allocates water for economic use according to its shadow price and accordingly equalizes the marginal shadow costs of water use across economic activities. We do not describe the institutional setup of such markets, nor the physical infrastructure and associated investment costs that may need to be in place for water markets to function properly, nor do we take transaction costs into account. In addition, water markets in our analysis should be considered more as a yearly market for water use rights (for example, in the form of an auction) rather than as a spot market to satisfy immediate short-term water use needs. Without water markets, shadow prices will, as a rule, differ across different economic activities and different locations.

1. *No water market*

In the no-water-market scenario, there is no possibility for exchanging water through a water market between sectors. That is, the water volume endowment is made immobile. This scenario is even more restrictive than the scenario calculated by models 1 and 2, as in those scenarios the water endowment was imperfectly mobile between crops so there was some reallocation of water between the crop sectors (with the same mobility as land).

2. *Two markets*

In the two-market scenario, two distinct water markets are specified. In this case, the agricultural water market supplies the agricultural sectors and the industry water market supplies the manufacturing and public water services sectors, but without possibilities of exchanging water between industry and agriculture. The water volume endowment is mobile, but a distinction is made between the water volume endowment that serves the industry water market and the water volume endowment that serves the agriculture water market.

3. *Single market*

In the single-market scenario, there is one water market that supplies all sectors. This is the most flexible option, where water can also be exchanged between sectors through the creation of a single market. The water volume endowment is mobile, and no distinction is made between the water volume endowments that serve the industrial or agricultural sectors.

4. *Single water market without public water services*

Perhaps for political or equity reasons, policy makers might find additional price increases for public water services unacceptable.Therefore, in this policy scenario,

Table 8.5 The scenarios and models examined in this chapter

Study and model used	Scenario name	Description
Study 1, (no economic model used)	Climate 1976, no adaptation	Climate impact on crop growth, no autonomous adaptation
1	Climate 1976, autonomous adaptation	Economic impact with autonomous adaptation of Warm+ agricultural conditions approximated by 1976 climate
2	Warm+ climate	Economic impact of Warm+ agricultural conditions
2	SW adaptation	Economic impact of Warm+ agricultural conditions with increased investment in surface water irrigation infrastructure as planned adaptation
3	No-water market	Economic impact of Warm+ climate conditions. Water is immobile
3	Two markets	Economic impact of Warm+ climate conditions. Water is perfectly mobile within (but not between) agriculture and manufacturing + public water services
3	Single market	Economic impact of Warm+ climate conditions. Water is perfectly mobile for all sectors
3	Single market without public water services	Economic impact of Warm+ climate conditions. Water is perfectly mobile for all sectors. Public water services is excluded

there is a single market for water-using sectors; however, the public water services sector does not participate. The public water services sector receives an 11% loss of summer water availability while the remaining summer water-using sectors collectively receive an 11% loss of summer water and reallocate the remaining water among themselves within a single market. By removing it from the water market, public water services in essence receive a subsidy such that the shadow price of water that it faces is lower than the market price of water. Table 8.5 gives a summary of the scenarios and models examined in this chapter.

Table 8.6 Percent change in the price and quantity of the agricultural activities from the first study

Scenario		Wheat	Cereal crops	Vegetables and fruits	Sugar beets	Other crops
Climate 1976, no adaptation	Price change	0	0	0	0	0
	Quantity change	−13.0	−13.0	−9.0	−26.0	−7.0
Climate 1976, autonomous adaptation	Price change	3.2	4.6	3.7	5.4	2.4
	Quantity change	−14.1	−6.8	−8.7	−3.5	−10.4

Source Includes data from Koopman et al. (2015), Table 4

8.5 Results

8.5.1 *Study 1*

Table 8.6 shows the results at the aggregated crop level in the Netherlands from the first study implemented in model 1. The yield shock from the warm dry climate leads to a reduced output of agricultural goods. In the "Climate 1976, no adaptation" scenario there are no economic feedbacks and so the crop prices are assumed to be constant. Further there is no reallocation of economic inputs labor, capital, land and water so the economic losses are the direct result of the reduction in crop yields.

In contrast, applying the equivalent shocks from the dry climate of 1976 to model 1 results in economic feedbacks as a response to the reduction of crop yields. Due to the price-inelastic demand for agricultural commodities, prices rise which will in turn induce some new supply from Dutch farmers. Imports into the Netherlands will also increase as a result of increased crop prices. Further farmers will optimize their production plans in accordance with the new physical and market conditions. In the first place, recognizing the physical water scarcity, they will economize on the use of water and substitute (irrigation) water for other factors of production to the extent possible. They will also change their cropping plans in accordance with the new scarcity induced market conditions.

The prices of some crops are less affected than those of others. The extent of the price increase of Dutch agricultural products in response to the higher costs of production depends on the ability of foreign imports to make up for lost domestic supply. For example, the price of sugar beets increases in particular (see Table 8.6). Due to its "sheltered" position on the domestic market (very little imports and exports) it will be more difficult for consumers of Dutch sugar beets to find new suppliers to replace them. This raises the price for sugar beets and so at the margin farmers will prioritize the production of sugar beets.

Despite the adaptation responses to the new market and water scarcity conditions, the rewards of factors of production (labor, capital) in agriculture will decrease. As a third adaptation response, some factors of production may, at the margin, seek employment in non-agricultural occupations. In equilibrium, this would lead to a general decline of wages and rentals in the economy. This leads to a decrease of income of the owners of labor and capital: the households (i.e., the consumers).

The other sectors of the economy, broadly mining, manufacturing and services, are faced with two opposing effects: first, a decrease of domestic demand because of the fall in real income of the consumers, and second, a fall in wages and capital rentals that decreases their costs of production. A profitable adaptation response would therefore be to resource the supply of their goods from the domestic market to foreign markets. This adaptation response is easier for those sectors that already sell a large share of their output abroad (manufacturing) than for those sectors that primarily service the domestic market (a large share of services). In total when only examining the yield losses at constant prices with no adaptation possibilities for farmers, the 1976 approximation of the Warm+ climate would result in a 1115 million U.S. dollar loss in value to Dutch agriculture in the no adaptation scenario. When accounting for rising crop prices as well as the reallocation of water, land and other inputs the total value losses to Dutch agriculture is 612 million U.S. dollars. Taken together the rest of the Dutch economy now also has a loss of 405 million U.S. dollars, as the cheaper labor and capital are not enough to make up for the drop in domestic demand.

From these results it can be surmised that economic autonomous adaption has several aspects that should be considered when assessing the impacts of climate change. A reduction in the supply of agricultural goods leads to an increase in price. This price increase leads relatively more production of Dutch agricultural goods than the situation without the price increase and also leads to more imports. Both responses to the price increase help to satisfy consumer demands for food. A reallocation of production factors among the crop sectors based on the relative price increases of the crops leads to a very different estimation of damage and loss from climate change than would occur if one were only to look at the yield losses from the crop models. Finally the non-agricultural aspects of the economy are affected as well. The movement of labor and capital from agriculture to the rest of the economy depresses wages and capital rentals everywhere. While this price reduction helps the non-agricultural sectors, the reduction of consumer income and increased agricultural prices negatively affect the non-agricultural sectors. Therefore in assessing the economic effects of climate change it useful to look more broadly than agriculture alone even if agriculture were to be the only sector directly impacted.

8.5.2 *Study 2*

The Warm+ climate scenario as calculated by the NHI and AGRICOM models results in a decrease of rain fed yields for most crops in the Netherlands (with the exception of wheat and cereal crops). The optimistic assumptions on available water withdrawals in the NHI result in an increase in the production of irrigated crops in the

Table 8.7 Percent change in price and quantity of agricultural activities in study 2

	Warm+ climate		SW adaptation	
	Quantity	Price	Quantity	Price
Wheat	4.03	−0.83	4.69	−0.96
Cereal crops	1.72	−1.10	1.99	−1.28
Vegetables and fruits	0.72	−0.29	1.60	−0.63
Sugar beets	0.01	−0.31	0.01	−0.45
Other crops	1.86	−0.39	3.31	−0.69

Netherlands in contrast to the crop model and data for studies 1 and 3. The shocks to the productivity of water and land in study 2 are shown in Table 8.4 and the results on the production of crops in the Netherlands as calculated by model 2 are shown in Table 8.7.

In the Warm+ climate scenario, which includes autonomous adaptation, even though the impact on total crop yields in study 2 is opposite of study 1 (i.e. higher yields instead of lower yields), the economic adaptation mechanisms are the same. The increase in the supply of Dutch agricultural goods naturally leads to a drop in price also shown in Table 8.7. This drop in price benefits consumers and all other economic sectors that rely on agricultural products for their inputs. Wheat increases the most and has a lower drop in price as compared to cereal crops even through both activities have the same increase in land productivity as shown in Table 8.4. Wheat is very homogeneous product with a correspondingly high Armington elasticity (Armington 1969) which allows the sector to increase its exports and sell the additional supply. Sugar beets is mostly rain-fed so even though it has a high increase in irrigation water productivity, the small productivity loss from rain fed land keeps the total increase of output very small. Even with the very small increase in output, the sugar beet sector has a relatively large reduction in price, this is because almost all of the sugar beets grown in the Netherlands are consumed by domestic markets which makes finding new avenues for exports relatively difficult. In Vegetables and Fruits and Other Crops by comparison a large proportion of domestic production is already exported (70 and 80% respectively) which makes it easier to increase exports rather than saturating the Dutch market, therefore there is relatively little decrease in the price of these products. The cereal crops sector has a large reduction in price in comparison to its increase in production (although not as large as sugar beets), indicating a difficulty increasing exports. This is a result of a relatively low share of exports compared with domestic production (35%) and a very low Armington, the lowest of all of the crop sectors.

The decline in the price of agricultural products also has an effect on the endowments. This is shown in Table 8.8. The marginal value product of surface water and land irrigation decrease slightly, while the marginal value product of the direct substitute ground water and land irrigation slightly increases. The marginal value product of rain fed land decreases substantially. Most of the rain fed land endowment is used

Table 8.8 Percent change in price for agricultural endowments in study 2

	Warm+ climate	SW-adaptation
Surface water	−0.88	−39.49
Ground water	0.59	−4.65
SW-land	−0.83	−0.60
GW-land	0.61	−4.63
Rain fed land	−9.00	−12.75
Labor	0.03	0.05
Capital	0.02	0.04

by Vegetables and Fruits and by Other Crops so even though the productivity of rain fed land increases for Wheat and Cereal Crops the average total productivity of rain fed land in the Netherlands decreases. We have assumed that similar products are grown in the rain fed and irrigated components of a single aggregated crop sector in so doing we assume that if rain fed land becomes less productive then the same crop can be grown on irrigated land without too much difficulty this is reflected in the very high elasticity of substitution between rain fed and irrigated land.

Almost all of irrigated agriculture in the Netherlands is Vegetables and Fruits and Other Crops. From Table 8.7 we see that the production of Other Crops has a larger increase as compared with Vegetables and Fruits. This in turn results in a higher demand for production factors for Other Crops as compared with Vegetables and Fruits. The surface water and land endowments are evenly distributed between the two crops however the groundwater and land endowments are primarily (about 80%) used by Other Crops. As a result the price for the groundwater and land endowments have a small price increase compared with the surface water endowments.

In the Warm+ climate scenario the price of labor and capital also increase. Demand for these endowments increases when the agricultural production increases but unlike the irrigated land endowments there is no productivity increase in capital and labor. Therefore capital and labor are the relatively scarce resources compared to the base data without the climate shock. The stock of capital and labor are very large however so the increase in wages and capital rentals is small.

Compared with the Warm+ Climate scenario, the investment in surface water infrastructure further increases the production of all agricultural activities. The production of Vegetables and Fruits and of Other Crops increase directly through an increase in the surface water endowment and it induces these two large agricultural activities to use less rain fed land which further reduces the price of rain fed land, shown in Table 8.8. Table 8.8 also shows that the reduction in the shadow price of the surface water endowment (due to the increase in supply) reduces the shadow price of ground water and land as well as this is the substitute production method of surface water irrigated agriculture. The price of labor and capital increase even further than in the Warm+ Climate scenario however this increase is not enough to offset the decrease in rain fed land rents and so Wheat and Cereal Crops can increase their output even further as their input endowments are on the whole less expensive.

The gains of the productivity increase went to the owners of labor and capital, these make up about 50% of the expenditure of Dutch farms. As we assume that labor and capital are mobile through the economy this wage increase benefits workers and investors in all sectors not only agriculture.

Rain fed farmers (or more specifically the owners of rain fed land) are on average worse off as the productivity of their farms decrease while the productivity of their competitors increases. At the margin, given soil conditions and the costs of transition, rain fed farmers might choose to grow wheat as the productivity of this crop is higher in the Warm+ climate. Wheat can be more easily exported and the alternative rain fed components of irrigated crops have a relatively difficult time competing in the Warm+ climate.

8.5.3 *Study 3*

Table 8.9 shows the percentage change in quantity of water use in the various water using sectors in the Netherlands in study 3 for all four different water market alternatives as calculated by model 3. The results all take place in the context of the Warm+ climate change scenario from the Delta Program Phase 3, which has more conservative estimates on water availability for irrigation than in study 2, resulting in a reduction in agricultural yields compared to the base year. In the no-water-market scenario, the supply of water falls by an equal percentage in all water-using sectors. In response to the drop in water supply, the shadow price of water rises by far the highest in the manufacturing sectors followed by the agricultural sectors and less in the public water services. This is shown in Table 8.10 which displays the percent change in the (shadow) price of water for each scenario relative to the percent change in the price of water in the single-market scenario. The underlying reason for the difference in the change in shadow price of water in response to the same percentage reduction in water is that the average value product of a cubic meter of water used in public water services is lower than in agriculture and much lower compared with manufacturing. This coupled with the Leontief production technology between water volumes and dedicated water capital leads to the marginal value product of water being much higher in manufacturing than in agriculture or public water services.

The larger economy-wide changes for the Netherlands resulting from the climate shock and the four water market configurations are shown in Table 8.11. In the no-water market scenario, GDP output has the largest loss of all the various water market configurations. Agriculture loses 2.19% of its output, by far the largest of all the water use categories. Public water services lose 0.54% of its output, comparatively little, while manufacturing actually gains overall from the water-scarce climate. This is not surprising as agriculture endures a loss of land productivity in addition to the loss of available water. Further, the various land and water endowments in agriculture account for approximately 16% of the total value of primary inputs in production for the irrigated activities and 60% of the rain fed activities (wheat and cereal crops). In contrast, the percentage of the industrial water composite in the value of primary inputs in the manufacturing sectors ranges between 4% (paper and pulp) and 0.3% (other manufacturing).

The large drop in the output of agriculture reduces demand for labor and capital services which in turn reduces their market prices (wages). The reduced price of labor and capital services more than makes up for the loss of water for manufacturing as a whole. Therefore, manufacturing faces more favorable economic conditions in the water-scarce climate.

In the two-market scenario, the water supply is reduced by the same amount as in the no water market; however, two market mechanisms now exist, respectively, between the agricultural sectors and the industrial water users. As a result, the price of physical water is equalized separately within the agricultural and the industrial water-using sectors. From Table 8.9, we see that water moves from the public water

Table 8.9 Percentage change in quantity of water use in study 3

	No water market	Two markets	Single market	Single market without public water services
Vegetables and fruits	−11	−10	−7	−11
Sugar beets	−11	−7	−5	−8
Other crops	−11	−12	−9	−13
Manufacturing[a]	−11	0	−1	−1
Public water services[a]	−11	−12	−15	−11

[a]We only consider the use of summer water
Source Modified from Koopman et al. (2017), Table 5

Table 8.10 The ratio of the percentage change in the price of water over the percentage change in price of water in the single-market scenario

	No water market	Two markets	Single market	Single market without public water services
Vegetables and fruits	2.2	2.0	1.0	2.4
Sugar beets	3.1	2.0	1.0	2.4
Other crops	1.8	2.0	1.0	2.4
Manufacturing[a]	16.5	0.8	1.0	2.4
Public water services[a]	0.8	0.8	1.0	0.8

[a]We only consider the use of summer water
Source Modified from Koopman et al. (2017), Table 6

Table 8.11 Economy-wide results for the Netherlands under the Warm+ climate change water scarcity scenario

		No-water market	Two markets	Single market	Single market without public water services
Percent change in yearly output	GDP	−0.024	−0.021	−0.021	−0.021
	Agriculture	−2.20	−2.25	−2.07	−2.33
	Manufacturing	0.04	0.06	0.05	0.07
	Public water services	−0.54	−0.59	−0.72	−0.54
Percent change in price	Capital	−0.17	−0.15	−0.15	−0.16
	Labor	−0.16	−0.15	−0.14	−0.15

Source Koopman et al. (2017), Table 7

services sector to the manufacturing sectors which have essentially no loss of water with respect to the benchmark equilibrium. Within the agricultural market, water moves from the Other Crops to the Vegetables and Fruits and Sugar Beet sectors. The reallocation of water in the two-market scenario follows the difference in water shadow price from the no-water-market scenario shown in Table 8.10. Table 8.10 also shows that the percent change in water price is substantially higher in the agricultural market than in the industry water market which gives further insight into how water will move in the single-market scenario. The aggregated economy-wide results from Table 8.11 show that in the two-market scenario, the GDP output improves compared with the no-water-market scenario, and access to more water allows the manufacturing sectors to further increase their output. More productive allocation of water also increases the demand for labor and capital which increases their price compared with the no-water-market scenario. This results in higher costs for agricultural inputs which further depress output. Sharing a water market with manufacturing causes public water services to pay a slightly higher price for its water in addition to a higher price for labor and capital than in the no-water market scenario which also causes a reduction in output.

In the single-market scenario, all water-using sectors participate in a unified market resulting in a single price for water across all uses. In this scenario, water moves from public water services into agriculture as well as into manufacturing. The agricultural sectors use more water than in the two-market scenario while manufacturing sectors use slightly less, as the water demand from agriculture drives up the price that the manufacturing sectors face for water volumes. Public water services use even less water than in the two-market scenario. The single-market scenario is the best outcome for agriculture, as agriculture has the highest output in the single-market scenario compared with all other scenarios. Manufacturing does slightly worse than in the two-market scenario, and public water services have the largest loss in the single-market scenario compared with all other scenarios, as it faces higher prices for water, labor, and capital than in the other market scenarios.

Finally, we examine the scenario where there is a single market for water, but public water services do not participate and so only suffer the direct loss of 11% of its raw water supply. From Table 8.6, we see that as the water from public water services is not available, water moves from the agricultural activities to the manufacturing sectors. On the whole, agriculture has less water available to it than in the no-water-market scenario, but the remaining water is more productively allocated across the agricultural activities. The more efficient allocation of water within agriculture is not enough to offset the additional loss of water. In terms of total output, this is the worst scenario for agriculture. Manufacturing gains the most from this scenario, as the depressed price of labor and capital more than compensate for the higher price for water compared with the single-market and two-market scenarios.

Table 8.11 shows that even though instituting a water market increases overall output, any particular instance of the three water market possibilities considered will have winners and losers from the perspective of sector output. The particular winners and losers can be seen by examining the shadow price of water for each sector in the no-water market scenario (Table 8.10). If a sector's water shadow price is lower than the shadow prices of the other possible participants in a market or lower than the eventual market price of water in any instance of market, then the sector will loose from creation of a market. On the other hand, any sector with a higher shadow price for water than the eventual market price will gain. Put another way, a sector with a water shadow price in the no-market scenario that is lower (higher) than the market price for water that would exist in the instance of a particular water market is essentially receiving a subsidy (tax). Creating that water market would remove the subsidy or tax.

Further it is worth noting, that even though the single water market scenario is preferable from the standpoint of economic efficiency there may be other objectives in designing a planned adaption response to climate change, for example maintaining similar sector output levels in the new water scarce climate as in the base year. Agriculture has the largest losses from the water scarce environment while the manufacturing sector actually increases production in all water market scenarios including the no market scenario. There are two main reasons for increase in manufacturing, the first is that the industry water composite is combined with capital (Fig. 8.3) with a relatively high elasticity of substitution (0.5 compared with 0.1 for agriculture). The value of the capital expenditure is also very large compared the value of the water composite making substitution of the water composite with other inputs comparatively easy compared with the other sectors. Further the manufacturing sectors can take advantage of the reduced prices for labor and capital which result from the reduction in demand from the agricultural sectors. However, even though the output of the manufacturing sectors is less dependent on water availability than the other sectors it does have the highest shadow value of water compared with agriculture and public water services i.e. the manufacturing sectors have the highest ratio of expenditure on the water composite per unit of water compared with the other sectors.

Therefore if manufacturing enters into a water market with any of the other sectors, water on average will be allocated to the manufacturing sectors from the other sectors in the market.

The two markets scenario is also negative for agricultural output compared with the no market scenario even though agriculture does not participate in a market with manufacturing and water is more efficiently allocated within agriculture. The increased output of manufacturing in the two water markets scenario increases the demands on labor and capital which raises their prices compared with the no water market scenario. This analysis points again to the notion that it is important to look beyond the elements and actors directly participating in an adaptation intervention when assessing its impact.

8.6 Discussion and Conclusion

This chapter explores a range of interpretations and applications of the water endowment in a computable general equilibrium model to assess adaptation mechanisms to water scarcity from climate change. Three similar studies of economic adaptation in the Netherlands were examined. The first study examined autonomous adaptation as response to climate induced water scarcity while the second and third studies examined respectively investments in irrigation infrastructure and the introduction of a market for water as planed adaptation. All three studies found that when assessing the economic impacts of climate change it is important to look at the larger economy wide effects including the sectors that are not directly affected by the aspect of climate change under examination. In the first study for example the rising prices for agricultural products already partially compensated agricultural producers for their loss of output although the price increases were not uniform per crop type. The non-agricultural sectors on the other hand were not directly affected by the hot dry climate in this analysis but they were negatively affected by the reduction in Dutch consumer income and the increased price of agricultural inputs. This insight is useful in deciding where and how much to target policies aiming to reduce negative climate impacts on particular sectors or how they might be compensated for losses due to climate change.

The same is true when estimating the effects of a planned adaptation policy as examined in studies 2 and 3. It is important to look beyond the effects on the sectors directly affected by the intervention. In the second study the increased investment in surface water irrigation infrastructure helped farmers with access to surface water to take full advantage of the warm climate to increase their output. However the reduced crop prices which accompanied the increase in total crop output also affected the rain fed farmers as well as irrigated farmers with only access to ground water. In the third study the implementation of water markets increased the economic efficiency of the allocation of water, the scarce resource, and increased overall economic output. However, any sector participating in a water market with the manufacturing sectors would further reduce their output. Even in the two market scenario when agriculture was not participating in a water market with the manufacturing sectors, the increased economically efficient allocation of water allowed manufacturing to further increase production and in so doing increase the demand for and price of other resources (labor and capital) that are also needed in agriculture. These economy wide affects make an strong argument for the use of CGE models in assessing the impacts of water scarcity from climate change.

By examining these studies together we also make a methodological point that even in a single country such as the Netherlands interpretations of the water endowment can vary significantly. This is particularly true when the water endowment represents both the value of the water itself and also all expertise, capital etc. required to utilize the water resulting in increased consumption of the crop and higher crop yields.

The water endowment can be widely interpreted to mean many things and changes in the quantity, value and productivity of the endowment can in turn be interpreted in many different ways. To make a link with physical water it is important to be explicit about what the water endowment represents and if there are shocks to the endowment, where these shocks come from. As the studies 1 and 2 show the quantity of the agricultural endowment could equally be interpreted as the quantity of water or the quantity of machines and expertise on the farm to utilize that water. In the later a change in the availability of physical water results in a change in the productivity of the water endowment rather than a change in the quantity.

By separating out physical water from dedicated water capital in the second study, there is more clarity in the interpretation of a change in physical water availability. However this approach required some strong assumptions about the dedicated water capital endowment, namely that it was sector specific, with no elasticity of substitution between water volumes and dedicated water capital. This last assumption was necessary as self-abstracted physical water volumes don't have a volumetric price or value in the Netherlands, so the value of water had to be tied explicitly to the value of another endowment In any case being explicit about the interpretation of the water endowment in the scenarios examined can bring clarity in the interpretation of the value of water in various economic uses during times of scarcity.

Looking at the reactions of various economic sectors to the same water scarcity conditions, we can see that the impact on the sector from water scarcity is not merely determined by the nature of the water endowment itself but many other parameters in the model as well. Whether climate change and irrigation water availability positively (studies 1 and 3) or negatively (study 2) affect crop yields in the Netherlands, it is the mobility of endowment factors capital, labor and water which often determine the mitigating effects internal to the country. If a sector has a deficit (or abundance) of the water/land endowment, the mobility determined the ease with which it can attract other inputs. All three studies assume perfect mobility of labor and capital across all sectors. This assumption is important in determining, both the extent of the autonomous adaptation response to climate change and also the extent to which various economic actors suffer (or benefit) from the adaptation responses. In studies 1 and 3 a contracting agricultural sector has less demand for labor and capital which reduces their price. The assumption of perfect mobility allows these endowments to be used as inputs for the other economic sectors which mitigates the drop in price for agricultural labor (for example) which although is beneficial for agricultural laborers does mean that the farm owners must pay more for labor than would be the case in more segmented factor markets. In study 2, the increase in land and water productivity means that capital and labor are the relatively scarce resources in the production of the crops. In a more segmented factor market then the prices of agricultural labor and capital would rise even further as it would be more difficult to attract these endowments from the non-agricultural sectors.

Similarly the Armington elasticity determines the feedbacks on the international market. The elasticity determines the ability of a product to expand into other international markets or the extent to which it can be replaced by similar products of international origin which in turn determines the extent to which prices rise (fall) when the production of a sector decreases (increases). For example wheat has a high Armington elasticity, so when Dutch production falls the price stays the same as consumers can easily switch to other suppliers so producers do not get compensated by higher prices in contrast to the vegetables and fruits producers with a lower elasticity. It is difficult for the consumers of Dutch vegetables and fruits to switch to another supplier so the prices rise as a result. However when production increases the wheat farmers can significantly increase their exports while the vegetables and fruits producers can only do so marginally, therefore the price drops for the vegetables and fruits producers while it stays the same for wheat.

The methodology used in these studies has as usual some limitations. For instance, we assume the absence of transaction costs in any water exchange and furthermore, that sufficient infrastructure exists such that additional water can be abstracted at the point of use in any exchange of water abstraction permits. This assumption would overestimate the benefits of water markets. On the other hand, we assume a common shadow price for water in the benchmark equilibrium. If water abstraction is charged by volume, it is much more likely that agriculture will be charged water at a lower price than industry or public water services; if the baseline already contains this imbalance, then a water market would provide additional benefits not captured by this study. Further, in the extensions of the GTAP model used in this chapter, many of the behavioral parameters are distinguished by economic sector but not by region. This is also true for some parameters important to this chapter. Such as the elasticity of substitution between primary inputs (i.e. labor capital and land), the Armington elasticity of substitution between domestic and imported goods, and the elasticity of transformation (i.e. the mobility) of the endowments labor, capital and elements of the land and water nests. Estimating these parameters with local data might have improved the accuracy of some of the specific results in the studies considered in this chapter. The main trends and most importantly the economic mechanisms driving these trends would however remain unaltered.

References

Armington PS (1969) A theory of demand for products distinguished by place of production. Staff Papers-Int Monetary Fund 16:159–178

Berck P, Robinson S, Goldman GE (1990) The use of computable general equilibrium models to assess water policies. Working paper 545, Department of Agricultural and Resource Economics, University of California at Berkeley, Berkeley, CA

Briscoe J (2005) Water as an economic good. In: Brouwer R, Pearce D (eds) Cost-benefit analysis and water resources management. Edward Elgar, Cheltenham, pp 46–70

Brouwer R, Hofkes M (2008) Integrated hydro-economic modelling: approaches, key issues and future research directions. Ecol Econ 66:16–22

Brouwer R, Hofkes M, Linderhof V (2008) General equilibrium modelling of the direct and indirect economic impacts of water quality improvements in the Netherlands at national and river basin scale. Ecol Econ 66:127–140

Calzadilla A, Rehdanz K, Tol RSJ (2010) The economic impact of more sustainable water use in agriculture: a computable general equilibrium analysis. J Hydrology 384:292–305

Calzadilla A, Rehdanz K, Tol RSJ (2011a) The GTAP-W model: accounting for water use in agriculture. Kiel Working Papers 1745

Calzadilla A, Rehdanz K, Tol RSJ (2011b) Water scarcity and the impact of improved irrigation management: a computable general equilibrium analysis. Agricultural Econ 42:305–323

Collins R, Kristensen P, Thyssen N (2009) Water resources across Europe-confronting water scarcity and drought. EEA Report No. 2/2009, Office for Official Publications of the European Communities

Coumou D, Di Capua G, Vavrus S, Wang L, Wang S (2018) The influence of Arctic amplification on mid-latitude summer circulation. Nat Comm 9:2959

Dellink R, Brouwer R, Linderhof V, Stone K (2011) Bio-economic modeling of water quality improvements using a dynamic applied general equilibrium approach. Ecol Econ 71:63–79

Diao X, Roe T, Doukkali R (2005) Economy-wide gains from decentralized water allocation in a spatially heterogenous agricultural economy. Environ Develop Econ 10:249–269

Diao X, Dinar A, Roe T, Tsur Y (2008) A general equilibrium analysis of conjunctive ground and surface water use with an application to Morocco. Agric Econ 38:117–135

Dimaranan BV (ed) (2006) Global trade, assistance, and production: the GTAP 6 data base. Center for Global Trade Analysis, Purdue University, West Lafayette

Dupont DP, Renzetti S (2001) The role of water in manufacturing. Environ Resour Econ 18:411–432

Eurostat (2012) Annual detailed enterprise statistics on manufacturing subsections DA-DE and total manufacturing (NACE Rev. 1.1, D). http://epp.eurostat.ec.europa.eu/portal/page/portal/statistics/search_database Code:sbs_na_2a_dade

Eurostat (2014) Annual freshwater abstraction by source and sector. http://epp.eurostat.ec.europa.eu/portal/page/portal/statistics/search_database Code: env_wat_abs

Goodman DJ (2000) More reservoirs or transfers? A computable general equilibrium analysis of projected water shortages in the Arkansas river basin. J Agric Resource Econ 25:698–713

Gomez CM, Tirado D, Rey-Maquieira J (2004). Water exchanges versus water works: insights from a computable general equilibrium model for the Balearic Islands. Water Resour Res 40 (W10502)

Harou JJ, DE Pulido-VelazquezM Rosenberg, Medellin-Azuara J, Lund JR, Howitt RE (2009) Hydro-economic models: concepts, design, applications, and future prospects. J Hydrol 375:627–643

Hertel TW (1997) Global trade analysis: modeling and applications. Cambridge University Press, Cambridge, MA

Hoogewoud JC, Prinsen GF, Hunink JC, Veldhuizen AA, van der Bolt FJE, de Lange WJ (2013) Toetsingsrapportage NHI 3.0. Deltares, Delft

Hughes G, Chinowsky P, Strzepek K (2010) The costs of adaptation to climate change for water infrastructure in OECD countries. Util Poli 18:142–153

IPCC (2014) Annex XX: Glossary. In: Agard J, Schipper ELF, Birkmann J, Campos M, Dubeux C, Nojiri Y, Olsson L, Osman-Elasha B, Pelling M, Prather MJ, Rivera-Ferre MG, Ruppel OC, Sallenger A, Smith KR, St Clair AL, Mach KJ, Mastrandrea MD, Bilir TE, Barros VR, Field CB, Dokken DJ, Mastrandrea MD, Mach KJ, Bilir TE, Chatterjee M, Ebi KL, Estrada YO, Genova RC, Girma B, Kissel ES, Levy AN, MacCracken S, Mastrandrea PR, White LL (eds) Climate change 2014: impacts, adaptation, and vulnerability. Part B: regional aspects. Contribution of working group II to the fifth assessment report of the intergovernmental panel on climate change, p 1. Cambridge University Press, Cambridge, United Kingdom and New York, NY, USA

Klijn F, van Velzen E, ter Maat J, Hunink J (2012) Zoetwatervoorziening in Nederland Deltares-rapport, Delft

Koopman JFL, Kuik OJ, Tol RDJ, Brouwer R (2015) Water scarcity from climate change and adaptation response in an international river basin context. Climate Change Econ 6(1)

Koopman JFL, Kuik OJ, Tol RDJ, Brouwer R (2017) The potential of water markets to allocate water between industry, agriculture, and public water utilities as an adaptation mechanism to climate change. Mitig Adapt Strat Glob Change 22:325–347

Koopman JFL, Kuik OJ, van der Vat M, Hunink J, Brouwer R (2018) The economic impact of irrigation water scarcity from climate change: a CGE analysis distinguishing between surface and ground water. In Preparation.

Liu J, Hertel T, Taheripour F (2016) Analyzing future water scarcity in computable general equilibrium models. Water Econ Policy 2(4)

Ministerie van Verkeer en Waterstaat (2009) Waterbesluit—Besluit houdende regels met betrekking tot het beheer en gebruik van watersystemen (ontwerp)

Mulder HM, Veldhuizen AA (2017). AGRICOM 2.05 Theorie en gebuikershandleiding. Altera-rapport 2576d. Altera onderdeel van Wageningen UR, Wageningen

Ponce R, Bosello F, Giupponi C (2012) Integrating water resources into computable general equilibrium models—a survey. FEEM Working Paper 57, Fondazione Eni Enrico Matte, Milan

Renzetti S (1992) Estimating the structure of industrial water demands: the case of Canadian manufacturing. Land Econ 68:396–404

Rosegrant M, Cai X, Cline S (2002) World water and food to 2025: dealing with scarcity. International Food Policy Research Institute

Scharf D, Burke D, Villeneuve M, Leigh L (2002) Industrial water use, 1996. Minister of Public Works and Government Services Canada

Smit B, Burton I, Klein RJT, Wandel J (2000) An anatomy of adaptation to climate change and variability. Clim Change 45:223–251

te Linde AH (2007) Effect of climate change on the rivers Rhine and Meuse: applying the KNMI 2006 scenarios using the HBV model. Report Q4286. WL Delft Hydraulics, Delft

Teeples R, Glyer D (1987) Production functions for water delivery systems: analysis and estimation using dual cost function and implicit price specifications. Water Resour Res 23:765–773

van den Hurk B, Klein Tank A, Lenderink G, van Ulden A, Van Oldenborgh GJ, Katsman C, Van den Brink H, Keller F, Bessembinder J, Burgers G (2006) KNMI climate change scenarios 2006 for the Netherlands. KNMI De Bilt

Young RA, Haveman RH (1985) Economics of water resources: a survey. In: Kneese AV, Sweeney JL (eds) Handbook of natural resources and energy economics, vol II. Elsevier Science Publishers, Amsterdam

Zhu X, van Ierland E (2012) Economic modelling for water quantity and quality management: a welfare program approach. Water Resour Manag 26:2491–2511

Jason F. L. Koopman is an economist at Wageningen Economic Research. Jason is currently a Ph.D. candidate at the Vrije Universiteit, Amsterdam and his research interests lie in using computable general equilibrium (CGE) models to explore polices and trends in the areas of water, agriculture, bio-energy, and climate change mitigation at the national and global level. He is the author or co-author of several scholarly publications in those areas and has contributed to a number of projects from the European Commission and the Dutch Ministries.

Dr. Onno Kuik is a senior economist with more than 25 years of experience in the environmental economics domain. His main field of expertise is the use of computable general equilibrium (CGE) modelling to assess the economic and environmental impacts of energy and climate change mitigation and adaptation policies. Onno is the author and co-author of about 40 scholarly publications. He has worked in numerous research projects for the Organisation for Economic Co-operation and Development (OECD), the European Commission, the European Environment Agency, Dutch Parliament, European Parliament, and the Food and Agriculture Organization (FAO), in Europe and abroad.

Dr. Richard S. J. Tol MAE is a Professor of Economics at the University of Sussex and the Professor of the Economics of Climate Change at the Vrije Universiteit, Amsterdam. He specialises in the economics of energy, environment, and climate, and is interested in integrated assessment modelling. He is ranked among the 150 best economists in the world and the 50 most-cited climate scholars. He is the chief editor of *Energy Economics*. He has played an active role in international bodies such as the Stanford Energy Modeling Forum, the Intergovernmental Panel on Climate Change, the Global Trade Analysis Project, and the European Forum on Integrated Environmental Assessment.

Marnix van der Vat is a senior project manager and an expert in integrated water resources management with the Dutch research institute Deltares. He is trained both as a hydrologist with an M.Sc. in Physical Geography from Utrecht University, the Netherlands, and as an environmental economist with an MSc from the University of London. He graduated as an economist on optimization of management of a cascade of multi-use reservoirs on the Citarum River in Indonesia using hydro-economic modelling. Marnix has been involved in a large number of projects all over the world. He has managed the project Water Resources Inventory for the Republic of Azerbaijan focused on an analysis of surface and groundwater resources and their optimal use. He is working on the Ganga Basin Strategic Planning project in India integrating the water demand of different sectors such as hydropower, irrigation, water quality and ecology using a collaborative modelling approach for stakeholder involvement. Furthermore, he works for the World Bank on piloting methods for Climate Change Risk Assessment for investment projects. In the Netherlands, he has been leading the analysis of fresh water distribution underlying the Dutch Delta Program focusing on multi-sectoral strategic planning for climate adaptation. He is working now for the Dutch government on the integration of hydrology and economics in a risk based approach for the analysis of drought measures and water allocation. In the EU funded Horizon 2020 project IMPREX, he is working on the development and application of a generic hydro-economic model to analyse the economic impact of water allocation and drought measures.

Joachim Hunink is a researcher on geohydrology and is an expert in groundwater modelling. He works at the Dutch research institute Deltares. He is trained as a hydrologist with an M.sc. in Physical Geography from Utrecht University, the Netherlands. He has experience in local, regional modelling in area with a lot of data like the Netherlands and in area's with limited data availability. Joachim is involved at the Dutch Delta Program Freshwater for the last years and has been responsible for part of the calculations with the National Hydrological Model (LHM). He is also involved in quantifying the shortages and the effect on the different water users. Joachim is involved in several other groundwater projects on model development or system analysis in the Netherlands and abroad.

Dr. Roy Brouwer is Professor in the Department of Economics and Executive Director of the Water Institute at the University of Waterloo, Canada. He holds the University of Waterloo Research Chair in Water Resource Economics. Formerly, he was head of the Department of Environmental Economics at the Vrije Universiteit, Amsterdam (2009–2015) and chief economist in the Dutch Water Ministry (2000–2004). He is a visiting professor at the Swiss Federal Institute of Aquatic Science and Technology in Zürich, member of the External Advisory Board of the Vienna Technical University Doctoral Programme on Water Resource Systems, member of the International Scientific Advisory Committee of the Chinese Research Academy of Environmental Sciences, and Editor-in-Chief of the Elsevier journal Water Resources and Economics. His main research interests are in water resource economics, in particular water resource valuation, hydro-economic modelling and water policy instruments.

Chapter 9
The Economic Value of Seawater Desalination—The Case of Israel

Ruslana Rachel Palatnik

Abstract In the decades to come, seawater desalination is expected to be an important source of potable water in water-stressed countries bordering oceans and seas, in particular in regions with fast population growth where further decline of natural freshwater availability is expected due to climate change. Israel, being part of the "thirsty Middle East region", faces natural water scarcity that was exacerbated by the five-year drought started at 2013. This has led to an ever-growing reliance on alternative water sources, where seawater desalination has been the main source for addressing the shortages. Desalination plants currently provide about a quarter of the potable water supply with plans to expand the desalination capacity. In this chapter, I assess the economic value of seawater desalination which increases the water supply and diminishes the effects of natural freshwater shortages. The analysis employs the Israeli Computable General Equilibrium (CGE) model—IGEM, which estimates the benefit of desalination gained by reducing water shortages that cause an economic loss with impact on the entire Israeli economy. IGEM assesses the impact of water shortages on key economic indicators, where the shortage values are equivalent to the amounts of desalinated seawater—current and planned. The different qualities characterizing the five water types, which account for constraints associated with crop salinity-tolerance and food-safety regulations, are reflected in the model by the constant elasticity of substitution (CES) rates between different irrigation water types. The results reveal that the value of desalinated seawater due to its role in diminishing potable water shortages is about \$4 per m^3—much higher than its direct cost.

Keywords Water scarcity impacts · Desalination · Economic value · Computable General Equilibrium Modelling

---

R. R. Palatnik (✉)
Department of Economics and Management, and Sustainable Environmental and Economic Development research center (SEED), Yezreel Valley College, Afula, Israel
e-mail: rachelpa@yvc.ac.il

R. R. Palatnik
Natural Resources and Environmental Research Center (NRERC), University of Haifa, Haifa, Israel

G. Wittwer (ed.), *Economy-Wide Modeling of Water at Regional and Global Scales*, Advances in Applied General Equilibrium Modeling,
https://doi.org/10.1007/978-981-13-6101-2_9

9.1 Background

In this chapter, we use Israel as the case study for evaluating the economic value of desalination services made possible due to the access to the Mediterranean Sea. Natural water resources in the country are about to break historically low levels, following five years of hydrological drought (IWA 2018) started in 2013. This has led to an ever-growing reliance on alternative water sources, where seawater desalination has been the main source for addressing the shortages.

The water economy in Israel accounts for 2346 Million Cubic Meters (MCM) in 2016 (IWA 2017) with natural water sources accounting for only 39% of total supply (Fig. 9.1). The Israeli water economy entered a stage of energy intensive seawater desalination in 2005 with facilities built on the Mediterranean coast. Desalination plants, developed through public-private partnerships, currently provide about a quarter of the potable water supply (585 MCM, amounting to 70% of all of the water supplied to households) with plans to expand the desalination capacity to 750 MCM. The share of sectors in the demand side, and the water supply mix are presented in Fig. 9.1.

Yet, the recently published audit report, claims that the Israeli water sector is in crisis again, as the supply of the demand for water continually exhausts the natural water sources (State Controller 2018).

Israeli water sector has been developed historically under conditions of scarcity. Water management in Israel is considered efficient relative to other water abundant areas. Every drop is counted and priced according to average or marginal private (vs. social) cost. The growing dependency on seawater desalination, not only for meeting projected shortfalls due to increased demand, but also for improving drinking water quality, to counter the continuous deterioration in groundwater quality, was forecast in the late 1950s (Dreizin et al. 2008).

Yet, water management in Israel is mainly characterized by crisis management, rather than long-term planning (State Controller 2018). A decision to desalinate on a large scale was taken in 2000 as a result of Israel's growing water scarcity, which caused overexploitation of natural waters and rising salinity in the Northern Coastal Aquifer. The national goal was set to produce 750 MCM/year of desalinated water in 2020 (Government Decision No. 3533). Under a June 2008 cabinet resolution, there should already be six desalination plants. The State Comptroller's report (2018) states that the Water Authority did not take the necessary steps to meet the desalination target set by the government in 2008. Following several relatively rainy years, the perception of IWA was that already installed desalination capacity (of about 585 MCM) is more than enough to supply present and future water needs. As a result, a sixth desalination plant was not built (Globes 2018). The infrequent excess of supply in the years 2011–2012 was not used for natural water sources recovery as originally planned. Instead, water quotas for agriculture were increased, compromising the sustainability of natural water resources in Israel (State Controller 2018).

Figure 9.2 presents the water level of one of the main natural water sources in Israel—Lake Tiberias (Sea of Galilee). Once, this natural reservoir supplied up to

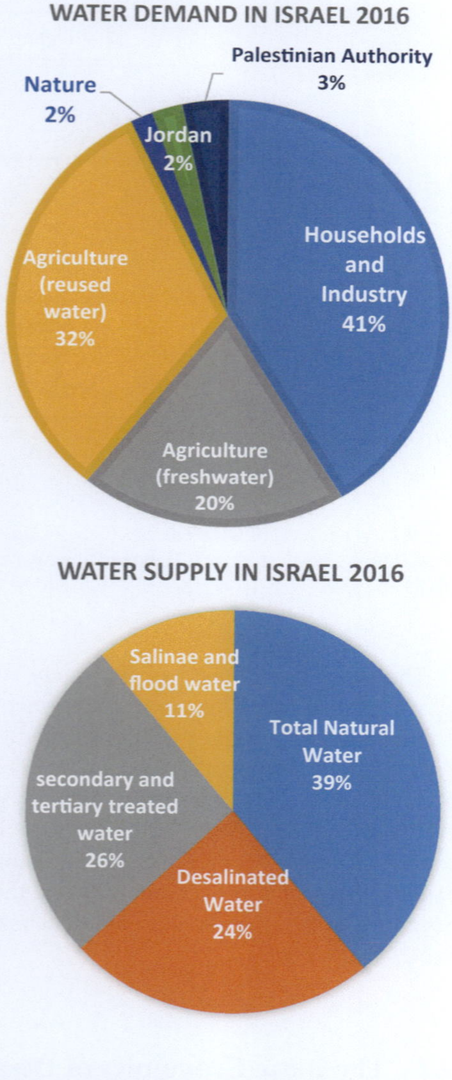

Fig. 9.1 Water economy, Israel 2016 (calculated based on IWA 2017)

one third of potable water in Israel. Continuous pumping in 2016–17, despite the prolonged drought, drove the water level of the Sea of Galilee below a critical level (the lower red line in Fig. 9.2), causing severe, and possibly irreversible, hydrological damage (State Controller 2018). In 2018, water supply from the Sea of Galilee dropped to zero.

At present, it is feared that continued drought is liable to disrupt home water supplies. The mismanagement and short foresight of water policy-makers, can be at least partly explained by undervaluation of natural water sources and of desalination.

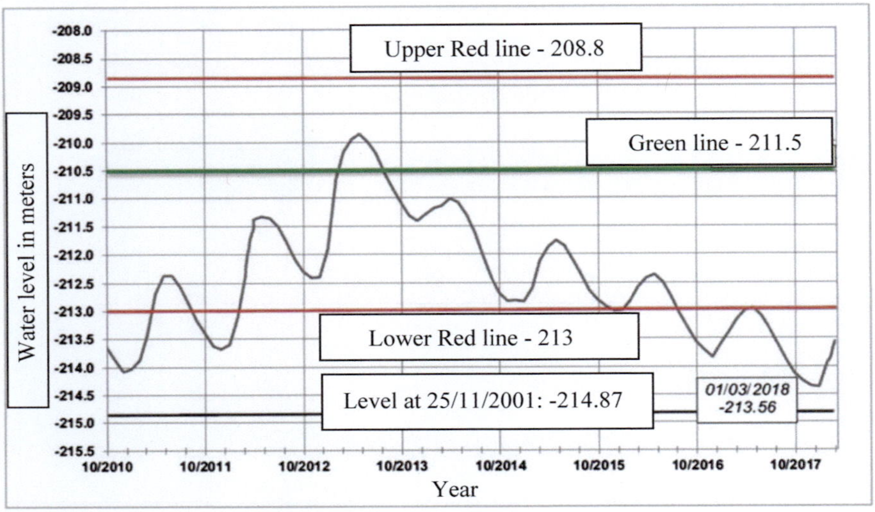

Fig. 9.2 Water level of Lake Tiberias (Sea of Galilee) in comparison to red lines, in meters. *Source* IWA (http://www.water.gov.il/Hebrew/Pages/Water-Authority-Info.aspx, Accessed 2 November 2018)

The novel contribution of this study is the assessment of the economic value of seawater desalination—a major service provided by the Mediterranean, which stabilizes the water supply and diminishes the effects of natural freshwater shortages. This is highly important for policy and decision makers as the economic value of marine services may be underestimated if based on their private cost or market price.

In what follows, Sect. 9.2 presents the basic economics of valuing desalination capacity. Section 9.3 introduces the economy-wide general equilibrium model with explicit representation of water sources, the data of Israel and the simulation basis. Section 9.4 summarizes the empirical results with some further discussion, and Sect. 9.5 concludes.

9.2 The Basic Economics of Desalination Value

Economic theory has some insight into the choice of the optimal size of a desalination project. While desalination plants provide many benefits through the supply of potable water to households, industry and agriculture, as well as allowing sustainable use of natural water resources, the full benefits of construction have often been ignored, both in the decision to build a plant and in the choice of the size of the water project. The externalities associated with overexploitation of natural waters are often ignored entirely, decreasing the perceived marginal benefit of desalination development.

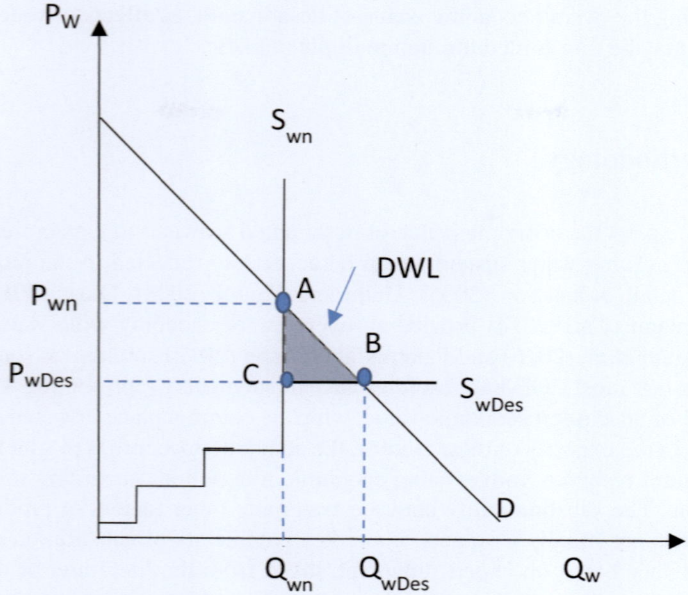

Fig. 9.3 Valuing the desalinated water

A simple static model depicts the forces that lead to underinvestment in projects such as desalination plants (Fig. 9.3). Let Q_W denote the capacity of a water sector. The aggregate water demand curve D reflects the marginal benefit of using water by all the sectors in the economy, as well as households, and neighbor countries.

If the aggregate water supply step function that aggregates natural water, treated and saline water sources is, SwDes and includes sufficient desalination capacity, then the equilibrium is reached at B, and the water price (PwDes) is equal to the marginal cost of water and to its marginal private benefit. This is the benchmark equilibrium of the current study.

Alternatively, if no/not enough desalination capacity is available—according to Master Plan scenarios (as explained below), then the supply of water is equal to the accumulated capacity of all water sources without desalination. In this equilibrium (A) the price (Pw) is defined according to the capacity constraint, being much higher than the marginal cost of the most expansive water source. We can refer to this difference between equilibrium price and the marginal cost as the scarcity rent of water. Thus, the value of desalination is the value of avoided "Deadweight Loss (DWL)" (triangle ABC).

As water is the major input to many economic activities as well as an important consumable, the true value of desalination is not only the partial equilibrium DWL avoided in the water market, but the avoided GDP loss that corresponds to the desalination level. A CGE model is the most appropriate methodological approach to estimating the overall economic value of desalination, as effects of water supply shortages may be transmitted through multiple markets.

9.3 Methodology

In order to assess the economic value of desalinated seawater in a macro-economic model, the existing water system has to be accurately reflected in the model with sufficient detail. Johansson (2005), Dudu and Chumi (2008), Dinar (2012, 2014) and Luckmann et al. (2014) provide an overview of economy-wide water-related studies. Baum et al. (2016) and Palatnik and Roson (2012) summarize some of the difficulties that most CGE-based studies have in adequately representing water, due to the lack of an explicit economic value, which is commonplace in water-abundant countries. In the majority of these studies, the adopted procedure is to split the value of this natural resource from existing economic transactions according to assumed proportions. The substitutability between water and other factors of production in CGE models are usually assumed, where key parameters of substitution elasticity rates are either based on expert judgment, taken from the literature, or used as a mechanism for calibrating the model. In most studies, only potable water is modeled (Horridge et al. 2005; Berrittella et al. 2007; Diao et al. 2008; Qureshi et al. 2012). In some of them, a distinction is made between irrigated and rain-fed agriculture, but there is usually only a single type of irrigation water (Dixon et al. 2011; Wittwer 2015; Hassan et al. 2008; Calzadilla et al. 2013). This modeling approach is not suitable for a water economy that relies on alternative water sources, such as desalinated, brackish water and treated wastewater since it does not reflect the constraints associated with utilizing low-quality water sources and the different costs entailed in producing the different water types.

Luckmann et al. (2014), Baum et al. (2016) and Yerushalmi (2018) are an exception. Using Israel as a case study, they model an economy with a representation for multiple water sources that differ for quality and costs. Such a detailed specification is particularly required in the case of Israel, where more than 75% of the wastewater generated by the urban sector is treated and diverted to agriculture, constituting approximately 50% of the country's irrigation water (Kislev 2011; Evgar 2018). While the irrigation share of treated wastewater in other regions is lower, e.g., 17% in Spain and only 6% in California (Sato et al. 2013), wastewater reuse is on the rise (Qureshi et al. 2012). In California, for instance, reuse has more than doubled from 1989 to 2009 (NWRI 2012). Thus, modeling variations in water scarcity in such countries should account for the substitution between freshwater and alternative sources.

The study employs the Israeli computable general equilibrium (CGE) model—IGEM (Palatnik 2009; Baum et al. 2016), to estimate the benefit of desalination, gained by limiting water shortages that affect the entire Israeli economy and cause an economic loss. As water shortages lead to increasing utilization of alternative water sources, the water sector in IGEM and the corresponding social accounting matrix

(SAM) were expanded to incorporate five water types: natural freshwater, desalinated seawater, brackish water and secondary- and tertiary-treated wastewaters. The following sub-section briefly describes the modelling setup.

9.3.1 IGEM—The Modelling Tool

The analysis of the macroeconomic aspects of desalination is performed with IGEM. This is a computable general equilibrium (CGE) model, developed for the Israeli economy (Palatnik et al. 2012; Davidovich et al. 2015). It provides a framework for examining the impacts of changes in the water system on the Israeli economy with special focus on the interrelation between the effects on sectors related to the water economy and changes in other sectors of the economy (Baum et al. 2016). The different qualities characterizing the five water types, which account for constraints associated with crop salinity-tolerance and food-safety regulations, are reflected in the model by the constant elasticity of substitution (CES) parameters between different irrigation water types. The CES parameters in this study are based on the evaluation provided by Baum et al. (2016). In addition, as tertiary- and secondary-treated water diverted to agriculture are produced from wastewater, whose quantities depend on the consumption level of potable water, the available quantities of treated wastewater were linked to the availability of potable water.

IGEM is a structural, static model of a small open economy, originally built with five energy commodities, fourteen other commodities, government, an investment agent and a representative household (Palatnik and Shechter 2008). As part of the research, the water sector in IGEM was expanded to obtain a more detailed and accurate representation of the water economy.

IGEM provides a full description of the economy with all its sectors, while representing the interrelations and interactions between sectors and the various agents that operate in the market. The principles of the model are based on the sound foundations of the neoclassical theory and reflect the interaction between different aspects of the economy such as taxation, economic development, welfare distribution, product allocation and trade (Hosoe et al. 2010).

The model adopts the standard assumption of market clearing, zero excess profits and a balanced budget for each agent. Commodity markets merge primary factors with producer outputs (Dinar 2014). The production processes in the various sectors are defined using functions that are based on constant elasticity of substitution (CES) between various production factors. In equilibrium, the aggregate supply of each good must be at least as great as the total intermediate and final demand. Producer supplies and demands are defined by producer activity levels and relative prices. Final demands are determined by market prices. There is a separation of activities from commodities which permits activities to produce multiple commodities, while any commodity may be produced by multiple activities. In addition, the model allows for heterogeneity between domestic and imported goods (Palatnik and Shechter 2010).

In terms of international trade, the model is a small open economy that cannot affect world prices. In this static model, the closure rule is to fix the balance of payments (i.e., export and import prices quoted in foreign currency are exogenous) because it would be difficult to interpret welfare when the current account deficit rises, for example, acting as a 'free gift' (Hosoe et al. 2010). In the long-run, the current account should balance on average.

In the present study, the exogenous shocks to the model are water shortages caused by increased water demand and/or decline in water supply. The results obtained from the model present the economic impact of the examined scenarios on the various sectors (changes in business level, prices, etc.) as well as on the economy as a whole (GDP). Thus, the economic value of the sea is revealed through the avoided GDP loss that would have been caused by water shortages in absence of the desalination capacity.

9.3.2 Data

IGEM assesses the impact of water shortages on key economic indicators, where the shortage values are equivalent to the amounts of desalinated seawater—current and planned. As water shortages lead to increasing utilization of alternative water sources, the water sector in IGEM and the corresponding social accounting matrix (SAM) were expanded to obtain a more detailed and accurate representation of the water economy. The model is calibrated to Israel's 2006 social accounting matrix (SAM), which was derived from the 2006 Use and Supply tables and the 2006 Satellite Account of Water in Israel (Israeli Central Bureau of Statistics—CBS 2011).

Introducing marginal water sources into IGEM required adjustment of SAM—a multi-sector dataset, recording and combining the transactions between different industries, consumers and government agents. The SAM contains information on 18 sectors of the economy including the water sector. The water sector in the SAM aggregates information on the value of water sales to the remaining 17 sectors, households, government, water import and export, as well as the value of the input factors purchased by the water sector from the other activities. To represent the marginal water sources, the original water sector in IGEM has been separated into five sectors—natural freshwater, desalinated freshwater, brackish water, secondary- and tertiary-treated wastewater.

To adjust the SAM we divided and allocated the aggregated data for the water sector across the various water types based on the Satellite Account of Water in Israel (CBS 2011). This report contains a comprehensive nationwide characterization of the economic value of the flows of the different types of water across the economic sectors. In addition, the characterization of the inputs used for the production of the different water types was performed based on Dreizin et al. (2008) and IWA (2011, 2017). This includes the assessment of the values of production factors such as energy and labor, as well as the values of inputs purchased from other economic sectors for the purpose of desalination and purification of effluents.

9.3.3 Scenarios

The scenarios fed into IGEM were derived from the Long Term National Master Plan for The Water Economy (IWA 2011)—hereafter referred to as "Master Plan". One of the main critics of the audit report (State Controller 2018) was that the Master Plan published in 2011 has not been approved by the government and therefore has not been fully implemented. However, the Master Plan provides an expert-based starting point for simulating the value of desalination. The future water balance included in the Master Plan presents the projected water supply against the expected needs based on data, assumptions and basic parameters of the water economy, coupled with design goals and water allocation policies. As to the future availability of natural freshwater, it takes into account a gradual reduction of 15% in the availability until 2050, (relative to the historic average between 1993–2008), particularly due to Climate Change.

The water balance presented in the Master Plan does not present an accurate forecast, but rather projected trends until 2050. This stems from the uncertainty with respect to events that influence both the supply and demand of water, including uncertainty related to future dependence of the water economy on international agreements (e.g. water export to Jordan and the Palestinian Authority) and decisions related to desalination. The Master Plan includes a probabilistic analysis of the gap between the forecasted demand and supply based on analyzing statistical variances of both.

The relative shortage in potable water expresses the expected shortage of potable water as a percentage of the overall needed quantity of potable water in that period, for given levels of desalination and reliability. The reliability levels in the IGEM model's scenarios correspond to the reliability levels in the Master Plan. These levels refer to the desired reliability of water supply to the economy and are derived from the probability of encountering a shortage in potable water—probability that stems from a statistical analysis of the gap between the projected future demand and anticipated future availability of natural potable water. Shortage of potable water at a reliability of 95% means that there is a probability of 5% that the shortage will be greater than the given value. Shortage at a reliability level of 100% means that the probability for a shortage greater than the given value is negligible. At the same time, it is important to note that even though the probability of having extreme values in all the events that influence the result in parallel, where each one is represented by a statistical distribution, is remote, this probability is nevertheless not zero. Hence, probabilistically what exists is not a reliability level of 100%, but only a reliability level that approaches 100%.

Table 9.1 summarizes the water shortage scenarios as calculated in the Master Plan. The scenarios list the expected shortage in potable water in millions of cubic meters a year (MMY) and as a percent of the total demanded quantity for the years 2020, 2030 and 2050, under two levels of desalination: the 2010 level of 280 MMY and a level of 750 MMY—the level approved by the government for implementation by 2020. The shortage in each scenario was calculated in the Master Plan under four desired water supply reliabilities (hereafter "reliability"): 75, 90, 95 and 100%.

Table 9.1 Water shortage scenarios

Scenario	Year	Desalination MMY	Natural fresh water MMY	Shortage of potable water (in % and MMY) at desired supply reliability level of potable water							
				75%		90%		95%		100%	
				%	MMY	%	MMY	%	MMY	%	MMY
1	2020	280	1,140	16	220	16	220	22	320	31	520
2	2020	750	1,140	–	–	–	–	–	–	4	50
3	2030	280	1,080	28	420	33	520	36	620	40	720
4	2030	750	1,080	–	–	4	50	12	150	19	250
5	2050	280	1,020	50	1020	52	1120	54	1220	56	1320
6	2050	750	1,020	35	550	39	650	42	750	45	850

Source Personal adaptation to Water Master Plan (IWA 2011)

The shortage and relative shortage in potable water were calculated as follows. The expected quantities of natural potable water (natural freshwater) in the years 2020, 2030 and 2050 for different level of supply reliability were taken from the Master Plan's basic scenario (IWA 2011, p 21) and appears in column 4 of Table 9.1. The overall shortage (not counting desalination) is taken from the Master Plan's chart (IWA 2011, p 23) which presents the result of a probabilistic analysis of the predicted gaps between supply and demand of natural potable water. Based on these results the shortage and relative shortage were calculated. The shortage was calculated by deducting the desalination level (either 280 or 750 MMY according to the scenario) from the overall shortage (for a given reliability level) whereas the relative shortage in percentages is calculated by the ratio between the shortage in MMY and the summation of the shortage in MMY and the quantity of available natural potable water.

9.4 Results and Discussion

IGEM reports, for each of the Master Plan scenarios, the percentage change in the value of key economic indicators relative to their value in the benchmark. In this article we focus on the most relevant indicators: the price of potable water, agriculture production, the price of agricultural products, and the GDP. The simulation results for 2020, 2030 and 2050 are presented in Table 9.2.

As the water economy is centrally managed by the IWA and the water prices are administratively determined, the relative increase in the price of potable water obtained in the simulation represents the price that ought to be set in order to reduce demand to match the available quantities. Alternatively, if the government decides to maintain the base-year water prices and limit consumption by setting quotas, the price increase would represent the shadow value of these quotas.

As expected, in all cases the price of potable water increases, agricultural production decreases, agricultural output prices rise and GDP declines. These changes increase over time as the shortage of potable water grows. Larger changes are obtained for higher water supply reliabilities; smaller changes are obtained for the higher desalination capacity of 750 MMY.

Changes in relative output prices alter the allocations of mobile capital, operator labor, dry land and irrigable land between activities. These factors follow a constant elasticity of transformation (CET) form.

To further study the value of desalination, IGEM was run under various desalination levels, assuming the reliability level of 90% for conditions projected for the year 2050. According to the Master Plan, natural water sources will supply about 1020 MCM, whereas the demand for potable water is projected at about 2420 MCM. Therefore, up to 1400 MCM of desalination might be required to meet the demand. It should be noted that following present practices, the increase in supply of treated and saline water is projected to be channeled to agriculture.

Table 9.2 Simulation results—changes in main indicators (%)

Year	Desalination MMY	Indicator	Desired supply reliability level of potable water			
			75%	90%	95%	100%
2020	280	Price of potable water	21.1	21.1	32.2	53.7
		Agricultural production	−3.2	−3.2	−4.7	−7.6
		Price of agricultural products	0.9	0.9	1.4	2.4
		GDP	−0.4	−0.4	−0.7	−1.0
	750		No Impact			Negligible negative impact
2030	280	Price of potable water	21.1	21.1	32.2	53.7
		Agricultural production	−3.2	−3.2	−4.7	−7.6
		Price of agricultural products	0.9	0.9	1.4	2.4
		GDP	−0.4	−0.4	−0.7	−1.0
	750	Price of potable water	Negligible impact	4.3	14.8	26.4
		Agricultural production		−0.6	−2.2	−3.9
		Price of agricultural products		0.2	0.7	1.2
		GDP		−0.1	−0.3	−0.5
2050	280	Price of potable water	21.1	21.1	32.2	53.7
		Agricultural production	−3.2	−3.2	−4.7	−7.6
		Price of agricultural products	0.9	0.9	1.4	2.4
		GDP	−0.4	−0.4	−0.7	−1.0
	750	Price of potable water	132.1	144.6	158.4	165.8
		Agricultural production	−16.4	−17.7	−19	−19.7
		Price of agricultural products	5.9	6.5	7.1	7.4
		GDP	−2.1	−2.3	−2.4	−2.5

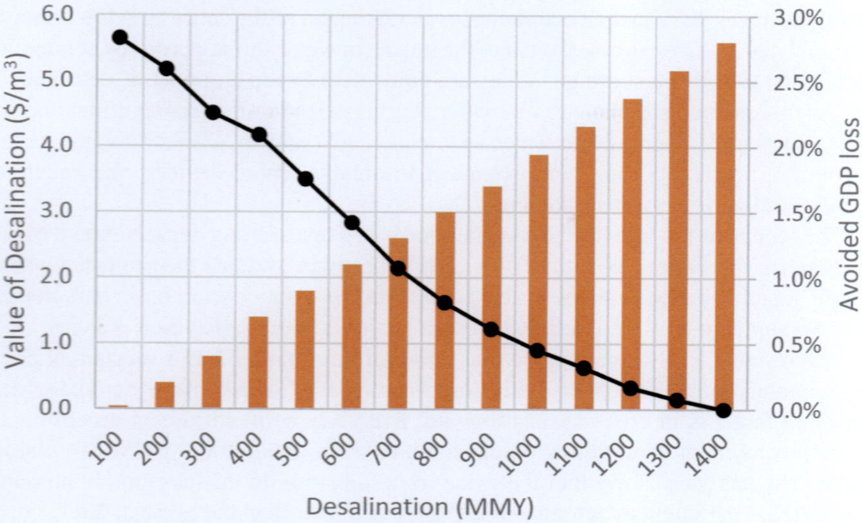

Fig. 9.4 The value of seawater desalination in terms of avoided GDP Loss

Figure 9.4 depicts the results. The right vertical axis shows the avoided GDP loss for every desalination level, and the left vertical axis shows the value of desalination in terms of \$ per m^3. Evidently, potable water shortage of 1400 MMY would cause about 2.7% loss of GDP (comparing to baseline values of 2006). Therefore, first 100 MCM of desalination capacity are valued as high as \$5.6 per m^3. As desalinated capacity increases, water shortage and corresponding GDP loss decline, diminishing the value of additional desalination capacity. Large fractions of the GDP decline are attributed to the impact on the value of agricultural and manufactured outputs. The agricultural sector suffers a larger decline than the manufacturing sector, but with GDP shares of 1.6 and 11.5% respectively, the impacts on both sectors equally contribute to the GDP decline.

The audit report states that as of January 2018, the Israeli water economy is short of about 1000 MCM (State Controller 2018). Therefore, the value of desalination is currently about four $/m^3$, comparing to the average direct cost of desalination of about 0.7 $/m^3$.

9.5 Conclusion

This chapter presents the economic value of seawater desalination, which stabilizes the water supply and diminishes the effects of natural freshwater shortages. The assessment was performed by employing the Israeli computable general equilibrium (CGE) model—IGEM, which estimates the benefit of desalination gained by avoiding water shortages that cause an economic loss with impact on the entire Israeli economy. The analysis includes an assessment of the impact of water shortages on key economic indicators, where the shortage values are equivalent to the amounts of desalinated seawater—current and planned. As water shortages lead to increasing utilization of alternative water sources, the water sector in IGEM and the corresponding social accounting matrix (SAM) were expanded to obtain a more detailed and accurate representation of the water economy.

The scenarios fed into IGEM were derived based on the Long Term National Master Plan for the Water Economy (IWA 2011). The plan presents the projected water supply against the expected needs based on data, assumptions and basic parameters of the water economy, coupled with design goals and water allocation policies.

The results clearly demonstrate that seawater desalination has a substantial economic value due to its role in diminishing natural freshwater shortages. The water sector in Israel is in crisis again following five years of hydrological drought and underdevelopment of desalination capacity, missing about 1000 MCM of potable water. The analysis shows that the value of desalination in this case might amount up to $US 4 per cubic meter, almost three times higher than the average direct cost. As high as this value might seem, it is lower than the social marginal benefit of desalination that includes not only the overall direct economic benefit, but also the external benefits for natural water sources. The inframarginal value is likely to be much higher because only the welfare distortions of water scarcity were assessed, while externalities were not addressed. The analysis demonstrates that the economic value of services provided by oceans and seas might be underestimated if evaluated based on their cost or market price. These findings have relevance for developing policies to support the efficient management of water sources.

Acknowledgements My appreciation is extended to Zvi Baum, Iddo Kan, and Mickey Rappaport-Rom, Mordechai Shechter and Ofira Ayalon for contributing to the study at various stages.

References

Baum Z, Palatnik R, Kan I, Rapaport-Rom M (2016) economic impacts of water scarcity under diverse water salinities. Water Ec Policy, Special issue on economics of salinity impacts and management

Berrittella M, Hoekstra AY, Rehdanz K, Roson R, Tol RSJ (2007) The economic impact of restricted water supply: a computable general equilibrium analysis. Water Res 41:1799–1813

Calzadilla A, Rehdanz K, Betts R, Falloon P, Wiltshire A, Tol RSJ (2013) Climate change impacts on global agriculture. Clim Change 120:357–374

CBS (Central Bureau of Statistics) (2011) Satellite account of water in Israel. http://www.cbs.gov.il/webpub/pub/text_page_eng.html?publ=72&CYear=2006&CMonth=10. Accessed 31 Oct 2018

Davidovich A, Palatnik RR, Ofira A, Shechter M (2015) An assessment of the impact of climate change on the insurance market: global and local analyses. In: Annual conference of the European association of environmental and resource

Diao X, Dinar A, Roe TL, Tsur Y (2008) A general equilibrium analysis of conjunctive ground and surface water use with an application to Morocco. Agric Ec 38:117–135

Dinar A (2012) Economy-wide implications of direct and indirect policy interventions in the water sector: lessons from recent work and future research needs. World Bank Policy Research Working Paper 6068, May

Dinar A (2014) Water and economy-wide policy interventions. Found Trends Microeconomics 10(2):85–165

Dreizin Y, Tenne A, Hoffman D (2008) Integrating large scale seawater desalination plants within Israel's water supply system. Desalination 220:132–149

Dudu H, Chumi S (2008) Economics of irrigation water management: a literature survey with focus on partial and general equilibrium models. World Bank Policy Research Working Paper 4556, March

Dixon P, Rimmer M, Wittwer G (2011) Saving the Southern Murray-Darling Basin: the Economic effects of a buyback of irrigation water. Econ Rec 87:153–168

Evgar Iddo (2018) Water sector in Israel. Knesset Research and Information Center. http://fs.knesset.gov.il/globaldocs/MMM/7d096829–0ff5-e711-80da-00155d0ad651/2_7d096829-0ff5-e711-80da-00155d0ad651_11_10526.pdf Accessed 8 Oct 2018

Globes (2018) Israel publishes tender for world's biggest desalination plant. Israel Business News, 29 Oct 2018

Government Secretariat, Government Decision No. 3533 on review of Israeli Water Sector and emergency plan for dealing with severe water crisis, 1 Jun 2008 [Hebrew]

Hassan R, Thurlow J, Roe T, Diao X, Chumi, S, Tsur Y (2008) Macro-micro feedback links of water management in South Africa. World Bank Policy Research Working Paper 4768, November

Hosoe N, Gasawa K, Hashimoto H (2010) textbook of computable general equilibrium modelling: programming and simulations. Palgrave Macmillan New York

Horridge M, Madden J, Wittwer G (2005) Using a highly disaggregated multi-regional single-country model to analyse the impacts of the 2002–03 drought on Australia. J Policy Model 27:285–308

IWA (Israeli Water Authority) (2011) Long-term national master plan for the water sector: policy document [Hebrew]. http://www.water.gov.il/Hebrew/Planning-and-Development/Planning/MasterPlan/DocLib4/PolicyDocument-jul-2011.pdf. Accessed 31 Oct 2018

IWA (Israeli Water Authority) (2017) Consumption of water in Israel 2016. Israeli Water Authority. http://www.water.gov.il/Hebrew/ProfessionalInfoAndData/Allocation-Consumption-and-production/20164/tericha%20lfie%20matrot%202016.pdf. Accessed 8 Oct 2018

IWA (Israeli Water Authority) (2018) The drougt continues! Press release, Isreali Water Authority March 22, 2018 [Hebrew]. http://www.water.gov.il/Hebrew/DocLib2/Hodaa_latikshoret_22032018.pdf. Accessed Sept 2018

Johansson RC (2005) Micro and macro-level approaches for assessing the value of irrigation water. World Bank Policy Research Working Paper 3778, December

Kislev Y (2011) The Water Economy of IsraelTaub center for social policy studies in Israel. Policy Paper No. 2011.15, November

Luckmann J, Grethe H, McDonald S, Orlov A, Siddig K (2014) An integrated economic model of multiple types and uses of water. Water Resour Res 50:3875–3892

National Water Research Institute (NWRI) (2012) Review of California's water recycling criteria for agricultural irrigation. Final Report. Prepared for the California Department of Public Health, Fountain Valley, CA

Palatnik RR (2009) Economics of climate policy impacts—applied general equilibrium analysis for Israel. VDM Verlag Dr. Muller Aktiengesellschaft & Co, KG, Saarbrucken, Germany

Palatnik RR, Shechter M (2008) Assessing the impact of greenhouse gas emission controls within the framework of a general equilibrium model of the Israeli Economy. Ec Q 55(4):545–573 [Hebrew]

Palatnik RR, Roson R (2012) Climate change assessment and agriculture in general equilibrium models: alternative modeling strategies. Clim Change 112:1085–1100

Palatnik RR, Faitelson H, Shechter M (2012) Israeli policy towards reaching Cancun pledge: a comparison of actions in plan and economically efficient measures. Finan Bus 2012:1

Palatnik RR, Shechter M (2010) The Israeli economy and potential post-kyoto targets. Israel Econ Rev 8:21–43
Qureshi ME, Proctor W, Young M, Wittwer G (2012) The economic impact of increased water demand in Australia: a computable general equilibrium analysis. Econ Pap 31:87–102
Sato T, Qadir M, Yamamoto S, Endo T, Zahoor A (2013) Global, regional, and country level need for data on wastewater generation, treatment, and use. Agric Water Manage 130:1–13
State Controller (2018) Annual Report No. 2018–969. ISSN 0334-9713 Jerusalem: State of Israel [Hebrew]
Wittwer G (2015) From almond shaming to water trading: CGE insights into managing California's drought. Centre of Policy Studies working paper G-258
Yerushalmi E (2018) Using water allocation in israel as a proxy for imputing the value of agricultural amenities. Ecol Econ 149:12–20

Dr. Ruslana Rachel Palatnik is the Head of the Department of Economics and Management and the director of SEED—Sustainable Environmental and Economic Development research center at the Yezreel Valley College, Israel, and the Senior Research Fellow at the Natural Resources and Environmental Research Center (NRERC), University of Haifa, Israel. Her research interests are climate change economics, environmental policy evaluation, energy and natural resource economics, and water management. She joined Fondazione Eni Enrico Mattei in 2007 in the Climate Change Modelling and Policy programme, where she was developing new integrated assessment modelling for climate-change impact and climate-change mitigation and adaptation policy evaluation focusing on agriculture, and land and water managing. In 2015–16 she was the visiting researcher at the Department of Agricultural and Resource Economics, University of California at Berkeley where she was analyzing the economic aspects of biofuels. She developed a CGE model of the Israeli economy and has been conducting climate change adaptation and mitigation policy analyses. She has evaluated the impact of water scarcity on the Israeli economy when water adaptation technologies include water recycling and various levels of desalination capacity. In addition, she carried out econometric research on external costs and economic incentives in solid waste management.

Correction to: The Expansion of Irrigated Agriculture in Brazil and Potential Regional Limitations

**Angel dos Santos Fachinelli Ferrarini,
Joaquim Bento de Souza Ferreira Filho, Santiago Vianna Cuadra,
Daniel de Castro Victoria and Mark Horridge**

Correction to:
Chapter 7 in: G. Wittwer (ed.), *Economy-Wide Modeling of Water at Regional and Global Scales*, Advances in Applied General Equilibrium Modeling, https://doi.org/10.1007/978-981-13-6101-2_7

In the original version of the book, the following correction has been incorporated in Chapter 7.

The sequence of the authors has been changed from 'Santiago Vianna Cuadra, Daniel de Castro Victoria, Angel de Santos Fachinelli Ferrarini, Joaquim Bento de Souza Ferreira Filho and Mark Horridge' to 'Angel dos Santos Fachinelli Ferrarini, Joaquim Bento de Souza Ferreira Filho, Santiago Vianna Cuadra, Daniel de Castro Victoria and Mark Horridge'.

The correction chapter and the book have been updated with the change.

The updated version of this chapter can be found at
https://doi.org/10.1007/978-981-13-6101-2_7

G. Wittwer (ed.), *Economy-Wide Modeling of Water at Regional and Global Scales*, Advances in Applied General Equilibrium Modeling,
https://doi.org/10.1007/978-981-13-6101-2_10

Index

G. Wittwer (ed.), *Economy-Wide Modeling of Water at Regional and Global Scales*, Advances in Applied General Equilibrium Modeling,
https://doi.org/10.1007/978-981-13-6101-2